THE ETHICS O

Power, Critique, Responsibility

Rainer Mühlhoff

BRISTOL
UNIVERSITY
PRESS

First published in Great Britain in 2025 by

Bristol University Press
University of Bristol
1–9 Old Park Hill
Bristol
BS2 8BB
UK
t: +44 (0)117 374 6645
e: bup-info@bristol.ac.uk

Details of international sales and distribution partners are available at bristoluniversitypress.co.uk

British Library Cataloguing in Publication Data
A catalogue record for this book is available from the British Library

ISBN 978-1-5292-4924-8 paperback
ISBN 978-1-5292-4925-5 ePub
ISBN 978-1-5292-4926-2 OA-Pdf

Cover design: Andy Ward
Front cover image: Better Images of AI/Anne Fehres, Luke Conroy and AI4Media

Contents

List of Figures and Tables

Figures

Tables

About the Author

Rainer Mühlhoff, Prof Dr, is a philosopher and mathematician leading the Ethics and Critical Theories of Artificial Intelligence research group at Osnabrück University, Germany. Grounded in continental philosophy, Rainer's work critically explores the societal impact of digital technologies through the lens of power relations and subjectivity. Integrating philosophy, media studies and computer science, Rainer's interdisciplinary research aims to develop a contemporary critical philosophy of the digital age.

Acknowledgements

This book initially began as a textbook version of my lecture 'Introduction to the Ethics of AI', at Osnabrück University, Germany, but soon evolved into a full monograph. My heartfelt thanks go to the first cohort of students who took this course in the Winter Term 2021–22, whose extraordinary positivity and engagement helped shape its early form. I also wish to thank the curators of the *Preis für gute Lehre* (Teaching Award) at Osnabrück University for awarding this lecture in 2022.

In writing this book, I benefited greatly from discussions with many people. For their inspiring feedback, I am deeply grateful to Anja Breljak, Gert Goeminne, Marte Henningsen, Daniela Hombach, Nora Lindemann, Hannah Ruschemeier, Paul Schütze, Jan Siebold and Annemarie Witschas. I am immensely thankful to my fantastic research team and my research colloquium for their careful scrutiny of earlier drafts. I am particularly appreciative of the tremendous support from my student assistants, Elena Herold and Anastasija Kocić. Special thanks go to Carl Hermann Mühlhoff for his photographic hunting of hoover nozzles.

I would like to thank Paul Stevens from Bristol University Press, whose encouragement played a decisive role in my decision to expand this work into a full monograph. My thanks also go to my copy editor, Scott Martingell, for his careful and professional work on the final typescript.

Last but not least, I thank my kids, as well as my family and friends in Berlin, for distracting me, supporting me and for their patience.

Introduction: What Does It Mean to 'Do' a Power-Aware Ethics of AI? A Note to Readers

This book aims to explore, in a manner both lucid and academically rigorous, some of the most pressing challenges posed by digitalisation and 'artificial intelligence' (AI) technology today. While my goal is to engage readers as critical scholars, I also seek to activate them as ethical and political agents. Hence, the book's 'mission' proceeds along two dimensions: it is scholarly in its method but extends beyond traditional scholarship in its ambition.

Concerning the first dimension, audiences in the Global North must come to terms with the reality that we are inevitably and involuntarily complicit in the workings of large data companies and their AI systems due to our everyday engagement with digital media. Let's be clear: the data we (the more privileged users of digital media) unintentionally generate fuel systems that enable the control, manipulation, exploitation and discrimination of people globally. Through a detailed examination of numerous everyday examples of digital, often AI-based technology, this book will illustrate this point in a way that personally addresses the reader. Building on these empirical insights, the book also puts forward ethical and conceptual arguments, introducing notions such as 'predictive privacy', 'control power' and 'collective responsibility'. These conceptual and normative analyses aim to inspire readers to participate in public, political and ethical discussions. True responsibility in relation to AI must be both a public and a political endeavour that transcends the prevailing political ethos of societies in the Global North, which typically revolves around individual benefits and interests in alignment with liberal notions of political subjectivity.

Second, the book argues that the traditional view of AI ethics as a branch of 'applied ethics', which largely ignores the dimension of power, is inadequate for addressing the unique challenges posed by AI. This approach has reached an impasse, caught between accusations of 'ethics washing' and an inability to confront the immense concentrations of power inherent to the capitalist-scientific-social complex of AI (Wagner, 2018; Bietti, 2020; Munn, 2022; Phan et al, 2022). By redirecting AI ethics to where it matters most, this book serves as a disciplinary intervention advocating for a different approach – one that is power-aware and structurally oriented, aiming to reshape ethics scholarship to meet the real, systemic and urgent challenges of AI.

To this end, my primary methodological conviction is that potent ethics must emerge from philosophical critique. By critique, I refer to the kinds of systemic and self-reflexive analysis that are rooted in the philosophical tradition, beginning with thinkers like Friedrich Nietzsche and Karl Marx, incorporating elements from Immanuel Kant and Baruch Spinoza, and finding its fullest expression in fields such as critical theory, feminism, postcolonial studies and 20th-century French poststructuralism. For me, critique as the basis of any viable ethics involves reconciling deconstructionist, normative and creative approaches to social and technological issues. Central to this approach are questions such as: What are the social conditions that allow this technology to exist, and conversely, what are the technological conditions that shape our societies and social lives? Who am I, in this digital age, as a cog in the machinery of techno-capitalist power? How did the current status quo become acceptable to us? And who is this 'us'? How can it be transformed into a 'we' that unites its diverse voices and perspectives into political action and meaningful change? What counter-narratives of technology, its limitations and desirable use, can we devise and popularise?

I will approach these questions through a broad two-step methodology in the ethics of AI (see also point 5 in the Manifesto at the end of this book): *critique first, normativity second*. Normative judgements and creative interventions are ineffective in addressing systemic harms unless they first question and dismantle the perceived naturalness, inevitability and givenness of the status quo. This begins with critically examining the very concept of AI, as every notion of AI carries with it implicit political and ethical assumptions. The idea that there could be an apolitical and ethically neutral concept of AI – or indeed of technology in general (see also Winner, 1980) – is an illusion. This is not an obstacle for ethics if transparency and awareness about this fact are admitted. In this spirit, Part I of the book deconstructs perspectives that present AI as purely technical objects, the neutral products of scientific endeavour and potential universal solutions to social problems. Instead, I reframe AI as a techno-politics deeply embedded within power structures and vested interests. Only after taking the essential steps of critically interrogating AI's entanglement with power should we[1] shift into a more normative or creative mode of discourse. By 'normative' I mean ethical discussions that focus on values, (collective) moral goods or specific kinds of harm and vulnerability relating to AI as sociotechnical systems. This is the approach taken in the book when addressing social inequalities and discrimination, using notions such as privacy, autonomy, self-determination and (collective) responsibility as key analytical tools.

As part of this two-step methodology, this book emphasises a forward-looking ethical engagement. Analytic approaches to isolated problems and corner cases in applied ethics – like the classic 'trolley problem' – inevitably fall short of addressing systemic issues and keep ethicists busy as functionaries within the current system. It is crucial to note that the forward-looking formulation and justification of specific values and principles, such as the concept of 'predictive privacy' (explored in Part II), rely significantly on epistemic and synthetic reasoning rather than on purely analytic methods. This process is not simply about weighing existing, widely recognised notions – like collective vs individualistic privacy – and then arguing in favour of one over the other. Instead, much of the task involves introducing a newly articulated value or moral good in ways that challenge supposedly common intuitions and core elements of established epistemes, as well as ethical and legal traditions. Creating space for new perceptions of injustice, values and moral harms that are not yet fully recognised, and developing concepts that enable these issues to be integrated into the epistemic frameworks of our debates, is *itself* an ethical endeavour. This ethical work is focused more on fostering epistemic transformation and creating new affordances for interventions than on traditional analytic philosophy.

The power-aware ethics of AI that I advocate for in this book is guided by a distinct philosophical ethos or attitude – one that could be seen as an exercise in virtue ethics. Central to this ethos is the understanding that scholarly contributions, including this book, are part of an ongoing (democratic, inclusive and vibrant) debate, rather than definitive treatises that aim to 'resolve' open questions or pass final, erudite judgement on ethical quibbles in some putative 'court of reason'. Accordingly, the ethical arguments I present here aim to convince sympathetic readers that certain developments in digital technology are harmful to society and individuals, rather than to decide or define once and for all what is right or wrong. Reaching consensus on what is good or bad requires cultivating a shared ethical and political sensibility and responsibility – something sorely lacking in our late-capitalist societies. This book aspires to contribute to that collective understanding, seeking to engage a sympathetic and receptive audience rather than winning over sceptics through compelling analytic argumentation. This kind of ethics aims to forge alliances among those who share my interpretation of – and deep concerns about – the current state of affairs.

Even more urgently, I should stress here that this kind of ethics leaves us all with extensive to-do lists, as it ultimately seeks its realisation in collective and political action. The primary arena for the creative work of ethics and responsibility-taking is collective and political action, where diverse stakeholders can unite, sharing their ideas, perspectives and vulnerabilities to envision and shape alternative futures with and for AI technology. The 'Manifesto for a Power-Aware Ethics of AI' in the concluding chapter of this book is intended to serve as an inspiration for precisely this kind of open and participatory quest.

Structure of the book

It is late in the year 2022. People from academic, political, business and media communities – as well as the wider public – are feverishly discussing the latest AI innovation, the chatbot system ChatGPT. Developed by the US-based start-up company OpenAI and backed with funding from Microsoft, this system has caught attention as a milestone in AI development due to its remarkable ability to engage in articulate, human–like conversations and provide stunningly well-versed answers even to complex questions across a wide range of knowledge domains (Lock, 2022). While large language models – the technology underlying conversational software agents such as ChatGPT – continue to attract widespread attention and fascination, their breakthrough has also given rise to a new wave of public discussion about their potential effects on society. These debates include questions of privacy and data protection (Ruschemeier, 2025 – forthcoming), corporate liability for generated content (Douglas, 2023), the impact on public discourse and politics (Kudina and de Boer, 2024), and this new technology's influence on fields such as education, law, journalism, programming and intellectual property, particularly in the fine arts (Sag, 2023; Dornis and Stober, 2024; Tanwar and Poply, 2024). In fact, it seems that ChatGPT has steadily been replacing the notorious 'self-driving car' and the obsession with 'trolley problems' in their role as standard examples that people bring up whenever it comes to an ethical questioning of AI (Paulo, 2023; Matzner, 2024: 152–6).

Attesting to the apparent ethical provocation of ChatGPT, in an unprecedented move, Italy's Data Protection Authority was quick to impose a temporary restriction on OpenAI in March 2023, prohibiting the processing of Italian users' personal data pending a more detailed investigation into the potential misuse of personal information and OpenAI's compliance with the EU's General Data Protection Regulation (GDPR) (Garante per la protezione dei dati personali, 2023). As a result, OpenAI temporarily blocked ChatGPT for Italian users until the dispute was resolved (Burgess, 2023; Ruschemeier, 2023a). Although the Italian Data Protection Authority's intervention was swiftly alleviated, ChatGPT sparked a heated discussion about the use of AI training data that are scraped from the internet, which some scholars doubt is legal (Ruschemeier, 2025 – forthcoming). OpenAI's large language model GPT-3, which was at the core of the chatbot service ChatGPT in 2022, was trained on 500 billion words' worth of text, gathered from multiple publicly available sources such as books, forums, websites, online encyclopaedia entries and more, all simply syphoned from the internet and social media sites (Cooper, 2023). In a similar fashion, competing large language models such as Google's Gemini, Meta's Llama and Anthrophic's Claude use scraped website data – for instance, from news outlets – for training. Likewise, in image generation, models like DALL-E, Midjourney and Stable Diffusion convert text prompts into visual outputs, based on extensive image datasets paired with descriptive texts.[2]

As all these examples show, exploiting rich troves of harvested data is the necessary foundation for generative AI systems to learn to simulate insightful conversations in natural languages. However, the data used for this end were not voluntarily provided for this purpose by the millions of internet users, book, weblog and website authors, journalists and so on who generated them, and neither were authors – individually or collectively – even informed about or asked to give their consent to this type of secondary use of their intellectual property. It is mainly due to ChatGPT that the mass processing of scraped data from the internet and social media is now being widely discussed as an ethical and legal issue as never before. Public debate is coming to grips with the fact that AI ventures rely on exploiting collective information resources. Slowly, AI business models are being increasingly recognised as a form of exploitation and appropriation of collective data troves.

Against the backdrop of a growing awareness of this issue, public debate is catching up with the discourse of this book. In fact, this book is about the problem of contemporary data-based AI systems fundamentally relying on the non-voluntary and exploitative participation of millions of people as data producers. As discussed in Part I of the book, the dependence of current AI – both as computational systems and as business models – on freely available or massively underpaid human-generated data is the very signature of the current 'third summer' (Kautz, 2022) of AI technology: a summer of AI systems that are designed to harvest and exploit the daily communication and information behaviour of users worldwide as a new form of labour. What is now being discussed in the wake of generative AI has thus always been the basis for the majority of the machine learning applications that have become popular everyday tools over the last 15 years, including scoring and rating algorithms, recommender systems, classifiers and so on. Making use of vast amounts of human-generated data is, as we will see in Chapter 1, a precondition of today's AI and the basis of its success, which is essential to the power of AI. Ranging from Google Search to image recognition (for example, Google Image Search, facial recognition on social media), speech recognition (Siri, Alexa, Google voice type) and translation tools (DeepL, Google Translate), these commercial AI technologies, accessed by billions of users daily, are built on vast amounts of user data, extracted unknowingly and without recompense.

Many scholarly contributions in the critical humanities and science and technology studies (STS) fields, particularly in the post-Marxist tradition, have tended to frame contemporary AI technology as enabled by the capture of (free or internet-based) labour (Terranova, 2004; Fuchs, 2010; Scholz, 2013; Fisher and Fuchs, 2015; Nixon, 2015). For example, in a recent contribution, Matteo Pasquinelli argues that 'the inner code of AI is constituted not by the imitation of biological intelligence but by the *intelligence of labour and social relations*' (Pasquinelli, 2023: 2, italics in original). This approach to AI is in stark contrast to the contemporary mainstream view of AI as the quest to 'solve intelligence', which is often seen as rooted in 'the secret logic of the mind' or the 'complex neural networks' of the brain (p 2). This scientistic understanding of AI, as Pasquinelli argues and as supported in this book, is 'a typical effect of ideology' (p 2). Based on this, Pasquinelli arrives at an important conclusion regarding the quest for some kind of working definition of AI in his post-Marxist reading: 'AI is a project to capture the knowledge expressed through individual and collective behaviours and encode it into algorithmic models to automate the most diverse tasks: from image recognition and object manipulation to language translation and decision-making' (Pasquinelli, 2023: 2).

The transition from imitation to capture is absolutely central to any critical and progressive analysis of the sociotechnical phenomenon of AI and its societal impact, as I have argued elsewhere (Mühlhoff, 2020c). AI technology today is much less about imitating, replacing or surpassing human/biological intelligence – although this is what many popular science discourses, corporate rhetorics or AI doomsday narratives want us to believe. Rather than replacing human intelligence, AI systems today rely on its *capture* (Chapter 2). They make use of human cognitive skills by immersing them in digital media and information networks. All the apps, gadgets and networked services we use every day constitute a global infrastructure of data harvesting and behavioural monitoring that forms the backbone of modern AI technology.

Following this interpretation, Chapter 2 introduces a philosophical approach that analyses AI systems as sociotechnical systems, serving as a crucial lens for viewing AI under the term 'Human-Aided AI'. Through this lens we see that what AI systems capture is not the 'knowledge' of humans directly, as Pasquinelli suggests, but more basically human cognitive skills as elicited in diverse human–computer interaction (HCI) contexts. The intimate and at times quite physical attachment of users to their digital devices such as wearables and smartphones (Kaerlein, 2018) gives rise to a specific form of power wielded by digital interfaces. This power takes hold of our bodies and brains directly and materially, not only through a layer of knowledge constitution and not only as part of a production activity that we should subsume under the catch-all term 'digital labour'.

Sourcing human cognitive skills is essential to contemporary AI algorithms, which thus *immerse* human cognisers as cognitive subunits in distributed, sociotechnical intelligence networks. So, to give a working definition of

AI for this book: *AI today is a project to capture the cognitive skills elicited in individual and collective behaviours by tacitly exploiting humans as subcomponents of algorithmic procedures.* All typical AI apparatuses today, which can automate the most diverse tasks ranging from image recognition to language translation, decision-making and even text, image or sound production, have this in common: they are *Human-Aided AI apparatuses* since they rely on continued access to human cognitive skills in the form of data input, (implicit) labelling, ranking, verification, moderation, training and so on (see examples in Chapter 2).

Subjectivity and power in contemporary AI

The concept of Human-Aided AI does not refer to a special kind of AI technology that exists alongside other kinds. It is rather a critical philosophical paradigm for analysing any AI system as a sociotechnical one. It is important to emphasise that the Human-Aided AI approach implicates *all users* of digital media services in this process, not just click-workers or those explicitly involved in digital labour (like labelling, content moderation or verification, for instance). In a networked society, it has become a standard business model to design the digital media services we use daily in such a way that they extract data and cognitive skills from our every interaction. Leveraging the data thus collected for AI purposes is a common strategy for monetising digital services, which are in themselves typically free to use. In Chapter 2, I will discuss this omnipresent participation and exploitation of users as a neocolonial form of extractivism, thereby aligning with the many critical contributions that make the same point (Thatcher et al, 2016; Coleman, 2019; Couldry and Mejias, 2019; Crawford, 2021). At the same time, the specific focus of this book will be on the non-geographical dimensions of digital colonialism and AI extractivism involving the exploitation of users worldwide and within societies even of the Global North. All of us, as users of search engines, navigation services, online shopping sites, dating apps, social media news feeds, content networks such as TikTok and YouTube, and language-based tools (DeepL, ChatGPT), are participating in AI networks under different forms of implicit or explicit engagement (Mühlhoff, 2020c).

Our participation is usually voluntary, often pleasurable and useful to us. It is therefore central to the approach of this book to analyse the *subjectivity* of using digital services – accompanied as they are by individual benefits, comfort, fun or social engagement – as a key ingredient in the way AI systems operate. User subjectivity is a crucial component of contemporary AI systems, understood as power apparatuses, and plays a central role in ensuring and perpetuating the availability of individuals worldwide as cognitive subagents. This constellation involves two related aspects: first, digital capitalism relies on effective strategies of user subjectification, such that individuals voluntarily and even eagerly assume the roles of data producers and cognitive subagents of large-scale intelligence networks (Chapter 3). Second, the power of AI apparatuses, in terms of both their 'intelligence' and their market power as businesses, derives from the myriad individual contributions of users that drive these very apparatuses.

To make sense of this, the main goal of the present book is to develop a *power-aware approach to AI* as the foundation of an ethics of AI. AI relies on the subtle integration of users into AI apparatuses, whereby their engagement is not merely an individual choice, but is as much a collective habitual disposition as it is a structural necessity for the functioning of these systems. This means that contemporary AI harnesses a form of power that works through the design of digital user interfaces of all kinds. This is a power that moulds and modulates individual and collective behaviour and subjectivity through digital mediation, thereby cultivating specific modes of reflexivity, sociality, labour and political participation that feel natural – even inevitable – while at the same time making people available for involuntarily cooperation in the attaining of algorithmic and capitalist ends (Chapter 4).

Contemporary AI as prediction power

One pertinent manifestation of the specific form of power enabled by contemporary AI apparatuses is what I call 'prediction power' in Part II of this book. As a technology that relies on the continued enmeshment of millions of users as data producers, the power behind AI not only entails the capture of past and present user behaviour. Rather, by means of its lateral comparison and learning from these data, this power also enables future behaviours and as-yet-unknown data points to be predicted. The ability to *predict* is thus an inherent capability of contemporary AI and one of AI's major applications today. In Chapter 5 I introduce the term 'predictive analytics' to refer to cases where AI systems are routinely deployed to estimate personal data such as age, gender, health risks, political interests, financial wealth or job performance based on readily available data that are gathered through social media or internet use.

Predictive AI is a technology for automatically sorting data subjects into buckets, and in most cases such predictive classifications are employed because the respective bucket influences how the operator of the system will treat the individual. Such purposeful *discrimination* is the existential basis of many of today's AI applications. Examples range from targeted advertising and personalised news feeds to differential pricing, hiring decisions or health scoring, as well as predictive border security controls and decisions on visa applications. As they are deliberately deployed to treat different people differently in ways that can be automatically applied to masses of individuals, predictive AI systems can easily contribute to patterns of unequal access to material and informational resources and opportunities that constitute lines of differentiation across societies. AI-based predictive capabilities, which give rise to what I refer to as 'prediction power' in Chapter 5, are therefore an essential manifestation of the informational power of AI technology in digital capitalism.

To be more precise, the capability to predict information about almost any individual comes in two varieties: it could refer to predictions of *future* developments such as security incidents or accidents, health risks and so on. But it could also refer to predictions of information that is already actualised but *unknown* to the predictor. This is evident, for instance, in predicting pregnancy or substance abuse on the part of a job applicant or supermarket customer as part of an assessment procedure (Duhigg, 2012; Morain et al, 2016). As I will argue in Chapter 6, in cases where unknown (and not necessarily future) attributes are predicted, the use of predictive AI capabilities could amount to a breach of the target individual's privacy. I will discuss how predictive breaches of privacy are often involved in contemporary forms of discrimination in the digital society – for instance, when automated hiring systems perform personality assessments for employment decisions (Hickman et al, 2022), potentially disadvantaging individuals with certain psychological traits or conditions. Under the title 'predictive privacy', I will argue that the prediction of unknown attributes constitutes a novel and largely unrecognised type of privacy violation that needs urgent ethical and regulatory attention (see also Mühlhoff, 2021, 2023b; Mühlhoff and Ruschemeier, 2024a).

It needs to be stressed that this ethical approach to privacy is a collective one that seeks to distance itself from the implicit liberal bias in the common understanding of privacy in many societies of the Global North. Predictive privacy refers to the social and political good of protecting society against the effects of an unregulated wielding of predictive power – and this power is given by the *potential* of data companies to predict private information about nearly *anyone* automatically and at scale. Not only protecting but even articulating the concept of predictive privacy comes with remarkable challenges derived from the fact that predictive AI systems operate precisely in those blind spots of the contemporary mainstream notion of 'privacy' in liberal societies.

First, the fact that predictive AI can estimate sensitive attributes that one has not disclosed in any other context goes against many people's expectations about what predictive AI can estimate. By deriving such information from less sensitive and often widely available data – for example, from social media or behavioural data – a user's privacy-related decision not to disclose these data can be circumvented. Data protection regulations, including the GDPR, do not effectively protect against this scenario (Mühlhoff and Ruschemeier, 2024a). A second, rather provocative challenge is what I call in Chapter 6 'collective enabling of predictive analytics': the training of predictive AI systems typically leverages data available from many *other* users. That is, predictions about a target individual are made possible by data collected about third parties (other individuals). It is often the lax privacy-related decisions of these third parties that enable the erosion of privacy of the target individual and of nearly *everyone* in society. This collective nature of predictive privacy and its infringement exploits a clear limitation of the individualistic understanding of privacy as the right to control one's own data. It shows that it is *not* in the control of the individual what information could be derived about them as long as other individuals can give away the relevant sensitive data at will and as long as data companies are allowed to train predictive AI models from these data (see Mühlhoff and Ruschemeier, 2024b). This problem leads to the third, and entirely counter-intuitive, challenge that predictive AI systems can generally be trained on anonymous data. This makes regulating the creation and application of predictive AI models particularly inaccessible to the current state of the art of regulatory frameworks, such as the GDPR, which addresses *personal* (and therefore non-anonymous) data.[3]

Along these lines I will also argue that prediction power comes with fundamental epistemological and methodological transformations that deeply impact many societal domains, including science, economics, medicine, public policy and administration. The technological and industrial complex of predictive AI is the driver behind a marked shift in epistemic culture from statistics to prediction, as I will discuss in Chapter 7. The relevance and scope of the issue of automated prediction is thus not only ethical and political, but also epistemological. Predictive AI forms a novel 'power–knowledge' nexus – a notion introduced by Michel Foucault (Foucault, 1978). AI-enabled prediction power is the most recent manifestation of power through a novel form of knowledge, constituting an informational power *asymmetry* between AI, data and platform companies on the one side, and societies and individuals on the other.

One of the main reasons for this book's strong focus on prediction power and its novel kinds of challenge to knowledge, culture, privacy and antidiscrimination is to point out that a truly valuable ethics of AI and its political implementation in the form of regulation has to be about the large-scale, society-wide effects of that technology. Predictive AI is a problem not primarily because it allows the privacy of a particular target individual to be breached or a specific person to be harmed, but rather because it allows these breaches and harms to occur simultaneously and automatically for a large proportion of users and citizens, leading to systematic differential treatment and discrimination of algorithmically determined 'groups'. The threat is to the mass, not just to the individual. Only from a collective viewpoint can we see that predictive AI is leading to large-scale differential treatment of individuals, which could go as far as structuring societies into virtual subgroups based on invisible differences when it comes to access to privileges and resources. This happens when the prices, offers and information we see on the web, the outcome of hiring or credit assessment procedures, and the workings of societal institutions such as welfare, education, youth protection, jurisdiction and the security services are adjusted according to the buckets into which we, as a result of prediction, are sorted.[4]

AI and social structures

The primary aim of this book is to address the most severe and pervasive effects of AI on societies as manifestations of power. To achieve this, it will prove useful to adopt a suitable conception of *social structures* and examine AI's role in establishing and stabilising such structures. I approach social structures as dynamic entities, referring to self-sustaining and stabilising patterns of differences, hierarchies and power relations that emerge from the interweaving of social, political and economic relations with AI technology in digital capitalism. The central claim is that the interplay between AI technology and society gives rise to virtual strata that permeate societies. Emerging from processes of computational and automated differentiation, these strata can manifest as patterns of discrimination, violence and vulnerability – for instance, in the form of racism, sexism, ableism, classism or as unequal access to material and informational resources and opportunities. Algorithmic strata can also manifest as different subtypes of subjectivity in the digital society – that is, as emergent alignments of self-perceptions, social practices, forms of pleasure and behaviours in relation to and as mediated by digital media services, as well as patterns of attachment and social relations, decision-making and narratives that emerge as a result of their interplay with AI technologies (Mühlhoff, 2020b).

The challenge in describing the socially structuring effects of AI technology is to think beyond the 'standard' grid of societal power hierarchies such as sexism, racism, classism, ableism and the like. As intersectional analyses of biases and unfair discrimination by AI have shown, some of the most egregious discriminatory effects of AI can only be detected and politicised by foregrounding the non-additive entanglements of discrimination along several of these axes (Angwin et al, 2016; Noble and Tynes, 2016; Buolamwini and Gebru, 2018; Chun, 2018; D'Ignazio and Klein, 2020). For instance, as Joy Buolamwini has been pointing out in a large body of her work, it is specifically dark-skinned women[5] who encounter the fate of non-recognition by face detection and facial analytics AI systems, such as those offered by Amazon, IBM, Microsoft and Face++. This mixed racialising and gendered bias cannot easily be seen just by testing the system's performance separately for the impact of gender and skin colour and hence has evaded attention for too long (Buolamwini and Gebru, 2018).

With regard to these types of constellation, the concept of *intersectionality* has been developed and popularised in critical race theory and Black feminism as an analytic framework that highlights the interlocking of different forms of oppression and privilege in the constituting of social identities (Crenshaw, 1989, 1991; Collins and Bilge, 2016; Noble and Tynes, 2016; Collins, 2019). The term is today often attributed to the legal scholar Kimberlé Crenshaw and her work in the late 1980s (see Collins, 2000). In a key 1989 piece, Crenshaw made reference to the historic legal case of DeGraffenreid v General Motors (USA, 1976), in which Emma DeGraffenreid and other Black female workers were fighting against mass lay-offs at the vehicle company General Motors that affected nearly all its Black female workers. While DeGraffenreid argued her case on the basis of 'compound discrimination' based on gender and race (Crenshaw, 1989), the court ruled instead that the case be argued separately on the grounds of race and gender discrimination. Since General Motors had also hired female office workers, the court found that the gender discrimination charge had been disproved; similarly, the hiring of Black male factory workers disproved the racial discrimination charge. The case was therefore dismissed.

This dismissal as the result of a mechanistic dimension–by–dimension analysis of the overall problem is very similar to how many software engineering teams are testing their AI systems for biases and potential discriminatory effects. As Buolamwini and Gebru pointed out, there is a crass underrepresentation of large demographic groups such as Black women in many benchmark datasets for AI testing and verification that form the gold standard widely used in the industry (Buolamwini and Gebru, 2018). This neglect makes intersectional biases in facial analytics AI systems an unsurprising outcome.

In the ethical approach of this book, it is essential to recognise that intersectionality is not simply a mechanical procedure for identifying combinatorial patterns of oppression and privilege. As Crenshaw herself explains, intersectionality is 'an analytic sensibility, a way of thinking about identity and its relationship to power' (Crenshaw, 2015). The key term here is 'sensibility', which indicates that intersectionality should be seen as an ethical and critical mindset. In the context of AI, the relevant axes of discrimination are not static – not even as complex combinations of established categories. Instead, patterns of algorithmic discrimination are 'high dimensional' and rapidly evolving, often defying traditional labels and their combinations (Mann and Matzner, 2019). As automated decisions by machine learning systems make use of hundreds or thousands of data fields, such as behavioural data and tracking data, going far beyond demographic attributes, and non-linear combinations thereof, we cannot expect discriminatory patterns of AI to follow named (or even nameable) and familiar paths. As a result, intersectionality calls for a deep ethical commitment that demands sensitive analysis and critical inquiry, embracing a forward-looking responsibility and a proactive search for hidden and emerging patterns of oppression and privilege.

The ethical sensibility of intersectionality as a critical perspective also underpins the approach to the ethics of social structuring effects advocated in this book. The invisible social strata – patterns and regularities of unequal treatment – produced by AI technology do not align neatly with familiar demographic categories. Therefore, describing and critiquing such social structures requires an epistemic *openness* to standpoints, voices and descriptions that make these structures visible. This ethos of recognising structuring effects is grounded in the ethos of intersectionality as an analytic sensibility.

In order to implement these insights at the core of a critical ethics of AI, I will outline a conceptual framework in Part III that makes social structuring understandable as processes of dynamically stabilising patterns of differentiation in the interplay of social relationships, AI technology and economic interests. To this end, in Chapter 8, I will start by making sense of contemporary machine learning AI in the legacy of cybernetics – the science of stewardship through feedback loops invented in the 20th century. This will enable us to rephrase the relationship between social structures and AI as one of *feedback loop control*: digital capitalism benefits from global structures of exploitation and inequality of different and always transitory kinds, and such social structures are dynamically stabilised and reinforced through the social effects of differential treatment facilitated by AI technology. The image of reality as represented by a specific AI model on the one hand, and the real social structures that materially shape life-worlds on the other, are each contingent, but both are in a relationship of reciprocal reinforcement. Breaking this relationship means addressing the power imbalance facilitated by AI technology at an aggregate level, by challenging the position of economic actors and their profit and efficiency metrics that drive the coupling of both sides of the equation.

Building on this, I will discuss in Chapters 9 and 10 that in addressing AI technology's role in creating and stabilising structures of social inequality, it is not enough to frame these effects as a problem of 'biases' in AI systems. Analyses of biases, for instance, have long pointed out that gender classifiers, differential insurance pricing, predictive policing and criminal recidivism-related scoring systems are racially biased in that ethnic minorities more often experience misclassification that leads to material disadvantages. Such findings, however, run the risk of ending up as free bug-reporting services for the negligent creators of the systems in question, unless they are integrated beforehand into an approach of ethics and critical philosophy of AI that insists on fundamentally questioning the very real purposes of such systems of automated governance and the social discourse that renders them acceptable.

Work on finding biases should alert us to the core questions of why systems for the automated unequal treatment of individuals are constructed and deployed in the first place, and how much of a hidden racist intention is inscribed in the very idea of algorithmic governance of populations that is inherent in any project that makes use of AI algorithms for these applications.

As part of this, ethics of AI must actively resist the idea that mitigating biases is a matter of fixing the bugs in an otherwise neutral and harmless technology. In particular, ethics of AI must resist the all too liberalist idea that the automated differential treatment of individuals could be ethically valid or socially desirable as long as no one is treated 'wrongly' – that is, based on erroneous predictions or misclassifications. The first reason for this, discussed in Chapter 9, is that the criterion 'wrong' is not even available as a valid descriptor when it comes to differentiations based on predictions of *future* actions, developments or attributes. In this regard, it is a commonly held but incorrect conviction that predictive AI's relation to reality is merely descriptive – similar to how statistics stands in a descriptive relationship to the reality it reflects. Compared to 20th-century-style statistical investigations (for example, in census-taking, in marketing, in medicine), AI models are usually much more directly and immediately coupled to automated actions in the world, with the 'learning phase' inseparable from the 'inference phase' (see Chapter 2, Note 2).

As a consequence of the direct coupling of probabilistic analysis and real-world action at a large scale, such predictions *create* the very differences they purport to merely foresee. Not hiring someone based on a predicted risk of substance abuse makes them statistically more likely to end up in a situation where they might really develop substance abuse; depriving customers who have been predicted to be less creditworthy makes their lives harder and more expensive and creates the very type of precarity that the system claims to predict. I will thus argue that there is a *performativity of predictive AI* – a risk of these systems producing the very realities they predict (see Chapter 8). This self-fulfilling prophecy effect exists precisely because AI systems are not detached from reality; they are not tools for merely analysing data and making inferences from statistics. Rather, these systems are sociotechnical ones that make use of humans (Human-Aided AI) *and* elicit feedback on them in the form of differential treatment and modulation of their everyday pathways in real time.

Furthermore, a critical theory of AI needs to be realistic enough to recognise that, usually, no one out there in the moulds of digital capitalism is really concerned with a 'correct', 'justified', 'explainable' treatment of each single person – as long as the treatment is more efficient and profitable *on average*. The rationale that drives automated decision-making is the striving for an efficient and profitable management of large numbers of people from an aggregate perspective. You see the advertisement for a customer service role, rather than the one for a product manager role, not because the algorithm has a deep understanding of your interests or skills, but because your data traces make you more similar to the people who have previously clicked on the first job ad rather than the second. Even if the prediction is wrong, it's still more efficient for the system to personalise ads in this way than not. In what I call the 'demise of explainability', we live in a world of large numbers and few reasons. In this world, the wrong or unjustified treatment of a marginal share of individuals is tolerated as long as there is an aggregate advantage in doing so. Demanding justification for the way you are treated is a lost cause because digital capitalism does not claim to treat anyone in a way that is true to their personality, but rather in a way that increases overall efficiency and profit for companies.

It follows, then, that in order to address social structuring, discrimination and inequalities as effects of AI, critique and ethics need to address the systemic level, where the intentions and capitalist rationale behind AI systems encourage a form of engineering geared towards optimising efficient and profitable management of masses of individuals and cases – and this often implies benefiting from social inequalities and patterns of exploitation. As this book argues, a cybernetic perspective on the interplay between predictive AI and social structures will help us to understand and criticise the systemic role of AI in the stabilising of the inequalities, social structures and global hierarchies of neocolonial exploitation: both reinforce each other in positive feedback loops. Predictive classifications are thus not descriptive statistical images of reality (as if in sociological theory), but are rather themselves agents of pre-emptive unequal treatment, which then become real and materialise as their performative effects.

Collective responsibility

Doing justice to the systemic impact of AI technology is a challenging task for classical ethics. As the discussion in this book lays out, the classical notions of moral harm and moral agency are fundamentally at odds with the situation at hand. This is because the typical and most pressing ethical issues with AI are not situations in which a distinct set of actors causes harm to a distinct set of moral patients. Rather, the most egregious social effects of AI technology are structural ones, which means that (1) they *potentially* affect large societal strata of people that are at the same time constituted with respect to being specifically vulnerable to these effects; (2) individual harms are often marginal, but aggregated harms can be egregious; and hence, (3) these effects are only visible from a collective viewpoint.

For these reasons, it seems far-fetched and overly theoretical to me to identify the moral behaviour of single agents, whether these are users or industrial actors, as the focal point of an effective ethics of AI scholarship.

Rather, the vector of my ethical discussion will be political activation and collective responsibility. In Chapter 11, I will turn to a more Aristotelian understanding of ethics as a *care for the polis*. 'Collective responsibility', in an understanding derived from Iris Marion Young and Hannah Arendt, will thus be the central pillar of my ethical approach. This responsibility, as Arendt points out, is necessarily forward-looking and political. Ethicists should not become mired in forensic questions of responsibility attribution for single cases of harm that were done, but should look at the *contingent* implications of innovations and decisions that unfold in the future. Citizens worldwide should have a say and raise their voice in the societal negotiation of futures with AI technology. Ethicists should not over-exaggerate the potential moral status of AI systems, which leads to an anthropomorphising of the technology and contributing to the industry-led and scientistic hype around AI that comes either in the guise of doomsday narratives or blue-sky promises of AI solutionism.

The moral agents in AI ethics are all of us and none of us. The moral agency is largely with the mass of people as users, political voters, workers and consumers. None of us alone can change anything, and yet the technology depends crucially on the availability of many of us as free subagents. This is why 'deleting Google' from your smartphone is not necessarily the outcome I seek from this discourse, but rather the activation of the public and a vibrant critical debate that pushes for stronger regulation and systemic political action to mitigate and prevent the most egregious societal effects of technological innovation. There is a huge gap in perception between our individual relationship to digital devices and services and their structural effects. We all know these devices and services directly and are interwoven with them in an almost intimate way. Public awareness and scrutiny of their societal effects as sociotechnical power apparatuses, however, lag far behind this level of individual entanglement. This is what we need to overcome if we are to engage in a genuinely public, ethical and political activation around AI ethics.

Ethics of AI is thus marked by a fundamental mismatch between the classic theoretical assumptions in ethics and the structural, collective nature of its objective. In this respect, ethics of AI has a lot in common with the problem of the imminent climate catastrophe. I would say the demise of classical ethics is a symbol for what are arguably the two major ethical-political challenges of our times: the global societal and structural impacts both of climate change and of digital media.[6] In both domains, it is the widely shared collective behaviour and habits that enable or drive devastating structural effects. It is the 'imperial mode of living' (Brand and Wissen, 2021) that implicitly shapes the behaviours of people in societies of the Global North that need to be scrutinised. It is not enough simply to blame individuals, because it is also industry and large organisations that exploit collective 'irresponsibility' in ways that lead to these effects. These effects have thus far been hard to communicate and difficult to discern as they are often marginal at the individual level, specifically when viewed from the limited perspective of the lifestyle bubbles of the more privileged in the Global North who may not know the people who are suffering from these effects. In both domains, ethical sensibility must be fostered and channelled towards political action that calls for better regulation. And in both domains, I am forced to admit that in 2024, a year marked by an upsurge of right-wing extremist forces in many societies of the Global North, as well as intense floods, heatwaves and wildfires worldwide – environmental crises that can be connected to climate change – the outlook is quite bleak. And yet, precisely because of this, we need to fight.

Intervening in the ethics of AI: a systematic delineation of the power-aware approach

The primary concern of this book is the systemic impact of AI technology as a form of structural power within contemporary digital media culture, shaping everything from social relations to politics. Notably, this structural perspective has received less attention within the emerging field of AI ethics, but has been more thoroughly explored in related disciplines such as STS, critical media studies, critical race theory and critiques of the digital economy. Traditional ethical frameworks often struggle to address the systemic effects of AI technologies, partly because they lack sensitivity to the material and sociotechnical situatedness of digital media and AI technology, and because their classical focus on moral agency, as well as on moral harm inflicted on moral patients, is inadequate for tackling the structural challenges posed by AI.

This book adopts an interdisciplinary approach to the ethics of AI, drawing on a broad spectrum of fields, including philosophy, media studies, STS, critical theory, digital anthropology, cybernetics and political science. My thinking and contribution to this field have been inspired by excellent, often empirically grounded works such as *Weapons of Math Destruction* (O'Neil, 2016), *Automating Inequality* (Eubanks, 2017), *Data Feminism* (D'Ignazio and Klein, 2020), *Algorithms of Oppression* (Noble, 2018), *Race After Technology* (Benjamin, 2019), *Gender Shades* (Buolamwini, 2017; Buolamwini and Gebru, 2018) and *Discriminating Data* (Chun and Barnett, 2021). More alternative theoretical approaches to ethics, most notably *Cloud Ethics* (Amoore, 2020), have encouraged me to engage with Foucault in the context of AI ethics. The concise introduction *AI Ethics* by Coeckelbergh (2020a) is excellent for its clarity on data-driven AI, while Virginia Dignum's *Responsible Artificial Intelligence* (Dignum, 2019) offers a refreshingly pragmatic approach, making key concepts accessible to engineers. Monumental works like *The Atlas of AI* (Crawford, 2021) and *The Age of Surveillance Capitalism* (Zuboff, 2019), as well as recent contributions such as *The Ordinal Society* (Fourcade and Healy, 2024) and more activist ones such as *Resisting AI* (McQuillan, 2022), have been invaluable inspirations. Shorter pieces and op-eds outside traditional academic publishing have been equally significant – for instance, Duhigg (2012), Angwin et al (2016), Grassegger and Krogerus (2016), Bogen (2019) and Goggin (2019). Another major influence on my approach comes from media studies and media philosophy (for example, Chun, 2017; Kaerlein, 2018; Matzner, 2024), particularly the works of Sybille Krämer (Krämer, 2015) and Alexander Galloway (Galloway, 2004; Galloway and Thacker, 2007), as well as the emerging subfield of interface critique (for example, Hadler and Haupt, 2016).

While these works mainly offer insights into the empirical and societal impacts of digital technology, my theoretical framework for AI ethics in this book is grounded in continental philosophy, particularly the critical traditions of poststructuralism and post-Marxism. Michel Foucault and Judith Butler have been central to my understanding of relational and productive power, as well as the concepts of subjectivity, subjectification and critique that underpin my analyses.[7] Baruch Spinoza's *Ethics* (Spinoza et al, 1994), along with Gilles Deleuze's interpretation of the former's work (Deleuze, 1988b, 1990), have deeply shaped my approach to philosophical enlightenment (see also Israel, 2001). As ethics is inherently enmeshed with questions of self-reflexivity and epistemology in Spinoza, a poststructuralist reading of his philosophy brings together the questions of critique, ethics and enlightenment (Saar, 2008; Mühlhoff, 2018c, 2020a) in a way that is immediately relevant to the present context of digital capitalism and AI. An early attempt to adapt poststructuralist conceptual tools for the critical analysis of power and subjectivity in the 'digital society' can be found in Breljak and Mühlhoff (2019). In this work, we sought to emphasise the distinct nature of reflexivity – a subject's relation to themselves, to others and to the world – under conditions of digital media, as well as the unprecedented forms of structural power that operate through us as users, contributing to the systemic effects of digital technology.

In terms of the Western[8] ethical tradition, the approach that most strongly resonates with my own work is that of virtue ethics. The similarity lies primarily in my focus on (1) larger sociotechnical constellations as potentially enabling or impeding what one might call a 'good life', (2) the subjectivity – including the moral character – of individuals involved in these sociotechnical constellations as a precondition for achieving a good life as a collective outcome, and (3) the mutually co-constituting relationship between subjectivity and the sociotechnical apparatus in question. In the back of my mind is Aristotle, who argues in the *Nicomachean Ethics* that ethics is not only a matter of education but also part of political science (Aristotle,

2011), showing that a good life can only be realised within a *society* that cultivates *character virtues*. As Shannon Vallor aptly puts it:

> Virtue ethics recognizes that we are not just actors facing isolated choices but subjects-in-formation constantly being acted upon by digital phenomena in ways that alter our own moral habits, dispositions, and capabilities. It dovetails neatly with the longstanding view in the philosophy of technology that human persons and societies are co-shaped by their technologies in morally and politically significant ways. (Vallor, 2024: 22)

Building on this key insight, this book develops a rich theory of 'subjects-in-formation' grounded in poststructuralism and a critical philosophy of power. This analysis enhances the traditional concept of character virtues, situating it within a deep scrutiny of the specifics of our digital condition. The character virtues most relevant in my approach to the ethics of AI are, as a consequence, critical questioning, enlightened discourse and collective responsibility.

At the same time, I have found that most existing approaches within the emerging subdiscipline of AI ethics do not fully address the systemic objectives of this book. First of all, I do not view the ethics of AI as merely a branch of applied ethics. If applied ethics involves transferring theses from normative ethics (such as utilitarianism, deontology or virtue ethics) into specialised fields, as is common in biomedical, business or environmental ethics, this approach falls short of delineating the true stakes that lie at the heart of AI's societal impact. While I acknowledge the success of biomedical ethics, and understand why some advocate for applying its procedural and institutional frameworks to AI ethics (Véliz, 2019), there are valid reasons to believe this would not be as straightforward or effective in the context of AI (Mittelstadt, 2019). Going beyond this quest, my aim is to show that digital technology and AI are far more than just tools or domains of application. Digital technology fundamentally shapes core practices of knowledge, social relations, labour, politics, culture, administration, finance and the economy. As such, it is not merely a subject for the application of ethical principles but a structural precondition that frames ethical deliberation and practice. It is precisely the implicit, contingent and often long-term or distributed effects of AI that elude the principles and approaches of applied ethics.

A similar dissatisfaction arises when we come to machine ethics as a subfield of AI ethics, which has sometimes also been termed 'machine morality'. These approaches follow the idea that machines could be 'artificial moral agents' and explore questions around the theory and design of the moral behaviour of 'superintelligent' agents (Moor, 2006; see Anderson and Anderson, 2011). Central issues in this field include the computational implementation of ethical principles, the capacities for moral judgement, the 'control problem' and the 'value alignment problem' – ensuring that the ('moral') decision-making of autonomous artificial agents aligns with human-created societal values (Müller, 2020; Wong and Simon, 2020). In my view, machine ethics fails to adequately address the inherently sociotechnical nature of contemporary AI systems. It falls into the trap of what I refer to as 'genius AI' (Chapter 1), which often underpins fantasies of 'AI takeover' and 'doomsday scenarios' (Witschas, 2024). The portrayal of AI systems as autonomous, potentially moral agents amounts to an anthropomorphising and animism (Dignum, 2019; Nyholm, 2023; Placani, 2024) of AI that is both misleading and detrimental to any critical scholarly or public discourse on the societal implications of AI. It also distracts from the pressing impacts AI already exerts in the present by projecting potential problems with AI into an imagined future. The infamous and overused trolley problem debate is yet another example of the irrelevance of machine ethics (Etzioni and Etzioni, 2017; Matzner, 2019) in this context.

Robot ethics, another subfield of AI ethics, similarly engages in the anthropomorphisation and attribution of autonomy to artificial agents, with a particular focus on material (robotic) entities (Lin et al, 2017; Gunkel, 2023). While the question of robots' moral agency is central to this approach, robot ethics goes a step further than machine ethics by also considering human moral obligations towards robots, akin to those we have towards other people or animals. This approach assigns moral status to machines, prompting some to advocate for the legal recognition of 'robot rights' (Gunkel, 2018). In my view, such arguments only deepen the misleading anthropomorphisation of AI systems, diverting attention from their true sociotechnical and systemic impacts and hampering critical discourse surrounding the societal implications of AI.

The attribution of moral agency to artificially intelligent machines remains a controversial topic within AI ethics. Scholars who advocate for this view are in the minority, but their work often garners significant attention in popular science and the media.[9] At the other end of the spectrum, there is a faction of AI ethicists who not only reject the idea of the moral agency of software systems, but also emphasise what they see as a fundamental ontological divide between humans and machines. They argue that because of the presumed unique ontological status of humans, moral agency lies solely with human users, operators and creators of technological artefacts. Artefacts, in turn, are seen as having an inherently *instrumental* relationship to humans. A prominent example of this view can be found in the growing field of digital humanism.

In their 2018 monograph, philosopher Julian Nida-Rümelin and film scholar Nathalie Weidenfeld present digital humanism as 'a new ethics for the age of artificial intelligence' (Nida-Rümelin and Weidenfeld, 2022: 4). The book critiques the depiction of AI and robotics in science fiction films as autonomous, anthropomorphic agents that could potentially threaten humanity, arguing that these portrayals fuel widespread societal misunderstandings about the nature of AI. This, according to the authors, forms the basis of what they term 'Silicon Valley ideology' (p 4). Philosophically, this ideology is based on two interconnected and complementary misconceptions: the mechanistic view of the human mind, which reduces the mind to material processes explainable by the laws of physics (examples include computational theories of mind), and the animistic view of machines, which attributes mental properties to algorithmic systems by projecting human qualities – such as emotions, intelligence, consciousness and moral agency – onto machines. (All of this has long been described and critiqued – for example, by Weizenbaum, 1976.)

Building on this undoubtedly accurate diagnosis, Nida-Rümelin and his colleagues proceed with an argument that is found not only in digital humanism but also in analytic philosophy and ethics, which makes it worth mentioning: textbook arguments from the philosophy of mind are used to 'falsify' the mechanistic view of the human mind and the animistic view of AI (such as determinism vs freedom and the qualia problem; see also Nida-Rümelin and Weidenfeld, 2022: 16; Nida-Rümelin and Staudacher, 2024: 24–5). This falsification is then interpreted as a confirmation of the humanist assumption of a fundamental ontological distinction between humans and machines. Hence, artefacts can never be moral agents. From this, they conclude that technology is ontologically confined to a passive role; artefacts (such as AI systems) are mere tools, subject only to correct or incorrect usage. This leads to the ethical imperative that we should adopt 'an instrumental attitude towards digitalisation' – and by a highly problematic implication, that we are also able to do so (Nida-Rümelin and Weidenfeld, 2022: 122; Nida-Rümelin and Staudacher, 2024: 18).

Certainly, AI systems are neither conscious nor sentient nor capable of moral action. However, this argument as a foundation for AI ethics is profoundly misguided and unsatisfactory. First, the debate over these questions has long been an obsession of analytic philosophy and ethics, distracting many theorists from the actual problems and real-world impacts of AI technology (Pistilli, 2022). Second, the leap from theoretical philosophical arguments to the instrumentalist view of (AI) technology as an ethical principle is a fallacy. Even though AI artefacts – lacking consciousness, sentience and existential grounding – cannot be moral agents, this does not mean that we can instrumentally control this technology and its social effects as we see fit. It is a fundamental insight of the philosophy of technology and STS that the effects and impacts of technology always go beyond mere instrumental use (see also Feenberg, 2008).

Third, the approach of rejecting popular anthropomorphising views of AI based on theoretical armchair philosophy arguments misses the point of why these views resonate so strongly in the world beyond academia – as corporate ideology, public imagination and persuasive narratives. Indeed, all the 'genius AI' imaginaries (Chapter 1) out there have an ontological effect: they are performative and potentially self-fulfilling as they shape public understanding of the risks and impacts of AI systems that already exist. Theoretical blindness to this fact overlooks how this view of AI is strategically exploited to achieve material effects, whose interests it serves and the significant societal harms it has already inflicted.

Against this background, the starting point of this book is straightforward: if there is a widespread belief that AI machines can think, feel, decide, evaluate, weigh options, take responsibility or act morally, there is a great chance that these machines *will inevitably be used for such purposes*, regardless of whether philosophers agree that they can. Through such use, the meanings of thinking, feeling, deciding and taking moral action will shift over time – they are by no means fixed concepts. My approach incorporates a degree of historical nominalism concerning what we consider to be 'human' qualities, and on these grounds I argue that the theoretical debate over whether AI can genuinely think or feel is practically, ethically, culturally, economically and politically irrelevant. What matters is understanding how it has become *acceptable* for so many people to delegate decisions, evaluations and creative work to machines. Whose interests does this delegation serve? What power-knowledge regimes are enabled by it, who benefits and who is disadvantaged? What are the societal consequences, from discrimination and exploitation to unprecedented levels of wealth accumulation? Rather than 'disproving' public and corporate beliefs with theoretical arguments, we need a philosophical critique in the fullest sense of the word: a material, sociotechnical, situated and power-aware analysis of why these beliefs are so pervasive and powerful, how we ourselves are implicated in these moulds of technological power and what alternative descriptions of AI could promote a more enlightened public discourse around this technology.[10]

Towards this end, the reflexive and simplistic dismissal of poststructuralism and STS in many analytic approaches to ethics is tragically counterproductive (see the 'Manifesto for a Power-Aware Ethics of AI' that concludes this book). For instance, the common trope of rejecting poststructuralism on the grounds that it allegedly assumes 'there is no truth and no universality' (Fuchs, 2022: 56) sets up the same problematic shortcut from ontology to ethics that I critiqued earlier: the idea behind poststructuralism is not to deny the existence of truth, but to critically examine a truth with respect to how it is constructed and how it functions within power-knowledge regimes. Seeking to avoid the 'God trick of seeing everything from nowhere' (Haraway, 1988) that is all too common in the erudite habitus of more analytical philosophical and ethical discourse, poststructuralist critique opens up a space for engaging a plurality of subjects in a joint questioning of how power structures and material interests are entangled with truths: Why do people believe this? Why has the world developed in a way that aligns around this (false, misguided and harmful) belief that is held as a truth? What is my own role in the power apparatus enabled through these beliefs? This historicising and self-critical questioning, rooted in the critical tradition of thinkers like Nietzsche, Foucault, Butler and Haraway, forms the foundation of the approach taken in this book.

However, this book also goes beyond the poststructuralist recipe and adds a layer on top: I maintain that we must also *dare to be normative* again (see the concluding Manifesto). As much as the strength of poststructuralist and STS approaches lies in their refusal to simply (once and for all) distinguish 'true' from 'false' from a presumed divine standpoint, they also enable us to engage situated subjects in a collective debate about what is 'good' or 'bad' for the kind of society we wish to live in. Ultimately, the critical ethos trained in deconstructing the power regimes entangled with modern technology must move towards making judgements and pushing for alternatives. We must be clear, assertive and vocal, not leaving the discourse to others who may be less occupied with self-critique. This is why I argue that any serious ethics of AI must follow the two-step methodology of critique first, normativity second. Skipping the critique and deconstruction phase and jumping straight into 'sorting out' what is right or wrong risks us becoming complicit with the status quo. On the other hand, getting trapped in endless self-questioning and deconstruction without moving towards judgement results in a purely theoretical and politically toothless exercise that ultimately fails to address the urgent problems we face today.

PART I

The Power of AI

1

What AI Are We Talking About?

Much of the direction any ethics of AI will take hinges on the presupposed understanding of the subject matter: What kind of technology are we actually talking about when we ethically debate AI? Since the mechanisation of intelligence and human thinking is a centuries-, perhaps millennia-old topic in science, philosophy, art and culture, which has always changed with the development of technology, there is hardly a universally valid definition of AI (see also Cave et al, 2020). However, in the context of this book my interest is entirely in our present – that is, in AI in the current decade, and maybe including the previous decade, as many of the contemporary techniques were built then. Hence I would like to pose the question of the definition of the term 'AI' differently: What kind of artificial intelligence is prevalent and socially impactful today? How *should* we understand AI today in order to enable the most relevant and urgent ethical debate? What terminology should we use to enable the most empowering debate about AI (see also Rehak, 2021)?

This question is an ethical one because our framing of AI already has ethical implications with regard to which aspects, applications and realities of AI are rendered visible and which ones are obscured. 'AI' is today used as a fuzzy term that can refer equally to technological and fictional creations, to actual achievements with real social consequences and to fantasy and science fiction, either utopian or dystopian (Witschas, 2024).

What characterises the contemporary use of the term 'AI', both in public discourse and in ethics of AI, is that it has something promising about it, something pointing to the future.[1] It is often difficult to tell which ideas associated with AI in news, political programmes, advertising, corporate communications and even philosophical debates are realistic or exaggerated, already implementable or merely wishful thinking. It is quite characteristic of the mainstream discourse on AI that it delves into imaginaries borrowed from science fiction films and books while many real and ethically troublesome manifestations of AI technology dwell in the blind spot of our cultural awareness of AI.

Popular examples of AI that are repeatedly presented in feature articles, science journalism and marketing include chatbots, self-driving cars, care robots, AI-driven surveillance technology and AIs that can play board games, such as the AlphaGo AI, which beat world Go champion Lee Sedol in 2016. In connection with self-driving cars, it is common to hear about the ethical dilemma of the so-called 'trolley problem' – a trope that comes first to many people's minds when they hear the term 'ethics of AI'. The trolley problem is a decision-making dilemma in which an autonomous vehicle finds itself in the fictitious situation of an inescapably imminent fatal accident, but it can still influence through its decision to steer the car in different directions which of various people or groups of people will be killed.[2] Robots and other embodied agents, both real and virtual (such as avatars), are also frequently discussed in AI ethics, with envisioned applications in eldercare, psychotherapy and even school tutoring.[3] Most recently, the release of the popular large language model–based chatbot system ChatGPT sparked a huge international debate on AI that covers many ethical and legal questions ranging from concerns about copyright infringement and plagiarism to its profound implications in education, as well as the dissemination of misinformation and its potential to manipulate democratic processes.

Human–computer interaction and 'genius AI'

As a general observation, most of the popular AI examples come with a specific and quite remarkable spatial framing of AI: the (artificial) intelligence is located *within* an artefact, often embodied either virtually or materially. That is, AI technology seems to be envisioned as manifesting in autonomous artefacts that interact with us *across* their outer boundaries (surface/interface), much as we interact with other humans, animals and technical appliances. While we never directly observe (artificial) intelligence, we are ready to *project* intelligence onto the innards of technological artefacts. This projection is based on manifestations of intelligence in the outer interaction of these agents.

This projective practice of ascribing (artificial) intelligence to machines (and locating it in their inside) was profoundly shaped in the second half of the 20th century by the notion of the Turing test. Ironically, Alan Turing himself stated that he considered the question 'Can a machine think?' 'too meaningless to deserve discussion' (Turing, 1950: 442). As a substitute, he suggested a kind of behavioural approach by suggesting what he called 'the imitation game', which has since become more well known as the Turing test. In this game, whether or not a machine is considered intelligent is judged based on whether that machine can be mistaken for a human being in its interactions. This test is less designed for embodied interaction and more for (written) verbal exchange. A certain, and remarkably narrow, media-technological framing is established as the set-up for this test: the machine to be tested for intelligence and a human being are each placed in single, confined rooms with a typewriter terminal connection to a third room, where a human 'interrogator' is located (Turing, 1950: 433–4). The interrogator can communicate through the typewriter interface with the two entities in the two rooms, without knowing who is the human test person and who is the machine. If the judge is unable to distinguish the machine in one room from the human in the other room, the machine passes the Turing test.[4]

The media-technological framing, which intended the test to be conducted through a then state-of-the-art terminal interface, was designed to ensure equality of opportunity, so to speak, between the machine and the human control person. The media-interactive channel with the human competitor needed to be restricted in its bandwidth for the competing machine to have a chance. Interestingly, the narrow framing of 'interaction' in the Turing test, and its associated concept of intelligence, did not face debate at the time. It appears that a specific cultural expectation of human–computer interaction (HCI) was widely shared in the media culture of the pre-Graphical User Interface (GUI) era, from the 1960s to the 1980s. This culture comprised two aspects: (1) the narrow channel of typewriter interaction between computer and user, and (2) the notion that an artificially intelligent computer's abilities are primarily 'mental' (as opposed to physical and embodied) and reside behind that interface in its interior, only to manifest implicitly in its written interaction. This conception of AI implies that intelligence is a virtual (insofar as it is invisible) capacity residing within a spatially bounded apparatus but manifesting itself in interactions with human beings across the boundary of that apparatus. As this boundary across which interaction takes place is a *media interface*, we see the significance of media technology in framing and shaping this conception of AI.

In the 20th century, a popular example that illustrates the impact of HCI on our notion of AI is automated board game play. Consider the historical chess computers of the 1980s. These computers played chess either on a screen or using a traditional chess board as an interface.[5] The system either displayed a virtual chess board on the screen or instructed an external operator on how to move the pieces on a physical chess board. In the famous chess match between Garry Kasparov and IBM's Deep Blue, a human operator digitised Kasparov's moves and replicated the computer's moves on the board.[6] The other player only experienced the computer's chess-playing 'intelligence' through that specific interface, the chess board. The computer's chess-playing intelligence had to fully manifest through that one channel – no conversations about chess with the computer, no books on chess strategy authored by the computer, and so on. This intelligence was bound to become evident as a marker of the computer's behaviour behind the interface and, as such, could not be directly observed. This media-theoretical logic of the interface mediating between the two 'intelligences' is deeply rooted in the 'classical' notion of intelligence at work here, which understood intelligence to be situated within the external confines of the intelligent entity as a material device.

The widespread cultural acceptance of this media-cultural imaginary is also demonstrated in Joseph Weizenbaum's observations about his ELIZA project of the mid-1960s, which is often hailed as the first chatbot. Originally created by Weizenbaum to illustrate the inherent limitations of human–machine communication, ELIZA was a chatbot that utilised pattern matching and formal substitution schemes to engage the user in a simple question-and-answer dialogue (Weizenbaum, 1966). After each user statement, the bot would generate a new, non-specific question by reusing some phrases from the previous statement. The project gained fame for one particular program, DOCTOR, which enabled the ELIZA bot to mimic a Carl Rogers–style psychotherapist (Rogers, 1951) by rephrasing ('parroting back') many of the patient's statements as questions and prompting the user to continue the conversation (see Figure 1.1). This particular program gained significant prominence in the 1960s, even beyond academic circles. In his influential book *Computer Power and Human Reason* of 1976, Weizenbaum recalls:

Figure 1.1: The 'DOCTOR' program on Weizenbaum's ELIZA bot of 1966 (reimplementation by Norbert Landsteiner, 2005)

```
Welcome to
                EEEEEE  LL      IIII  ZZZZZZ  AAAAA
                EE      LL       II       ZZ  AA   AA
                EEEEE   LL       II      ZZZ  AAAAAAA
                EE      LL       II      ZZ   AA   AA
                EEEEEE  LLLLLL  IIII  ZZZZZZ  AA   AA

    Eliza is a mock Rogerian psychotherapist.
    The original program was described by Joseph Weizenbaum in 1966.
    This implementation by Norbert Landsteiner 2005.


ELIZA: Is something troubling you ?
YOU:   Men are all alike.
ELIZA: What is the connection, do you suppose ?
YOU:   They're always bugging us about something or other.
ELIZA: Can you think of a specific example ?
YOU:   Well, my boyfriend made me come here.
ELIZA: Is it important to you that your boyfriend made you come here ?
YOU:   He says I'm depressed much of the time.
ELIZA: I am sorry to hear that you are depressed.
YOU:   It's true. I am unhappy.
ELIZA: Can you explain what made you unhappy ?
YOU:
```

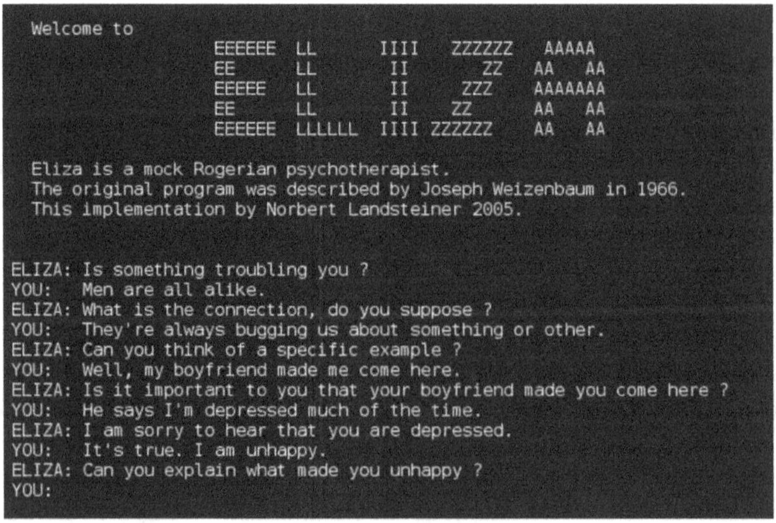

Source: https://commons.wikimedia.org/wiki/File:ELIZA_conversation.png

I was startled to see how quickly and how very deeply people conversing with DOCTOR became emotionally involved with the computer and how unequivocally they anthropomorphized it. Once my secretary, who had watched me work on the program for many months and therefore surely knew it to be merely a computer program, started conversing with it. After only a few interchanges with it, she asked me to leave the room. Another time, I suggested I might rig the system so that I could examine all conversations anyone had had with it, say, overnight. I was promptly bombarded with accusations that what I proposed amounted to spying on people's most intimate thoughts; clear evidence that people were conversing with the computer as if it were a person who could be appropriately and usefully addressed in intimate terms. (Weizenbaum, 1976: 6–7)

The behaviour of Weizenbaum's colleagues and employees shows how culturally ready and willing people were to project personal interest, empathy and intelligence onto an apparatus. Since the 1970s, this observation has been termed the 'ELIZA effect' (see Hofstadter, 1995), which broadly describes a cognitive bias of humans to attribute depth, meaning, emotions and empathy to interactions with technological systems, even when they are aware that these systems do not possess real understanding or emotions.

This historical excursion demonstrates how the classical Turing test notion of (artificial) intelligence, along with the culturally widespread illusion of the ELIZA effect, reinforces the perception of (artificial) intelligence as residing within a physically bounded entity and being expressed externally across an interface through its interactions with external agents. Henceforth, I will refer to this classical, Turing test–driven and locally circumscribed conception of AI as 'genius AI'. Genius AI always operates through the ascription or attribution of intelligence in a dyadic relationship – that is, it indirectly locates intelligence on the other side of an interactive interface. Since intelligence can never be directly seen in one's interaction partner, genius AI implies its existence within the other interactant based on how it shapes that interactant's behaviour. Of course, there is nothing inherently wrong with attributing intelligence to an interactant. However, what remains largely unquestioned in this paradigm is the presupposition that intelligence is an inner property, capacity or force possessed by each interactant – an assumption I call 'locationism'.

Related to this locationism, the cultural and imaginary constellation of genius AI is propelled by an age-old notion of an aspired likeness between humans and machines in terms of cognitive abilities. Genius AI emerges from a projection that identifies machine intelligence through its resemblance to how we conceive human intelligence. Importantly, however, this influence between the depictions of human and computer intelligence is bidirectional.

In the historical context of genius AI, as much as presumably in any other historical period, the concept of human intelligence itself has always been shaped by what computers (or machines) could accomplish.[7]

Ethics of AI beyond anthropomorphism

The cultural imaginary of genius AI cannot only be challenged in its appropriateness for describing contemporary AI technology; this framing of AI also has important ethical implications. Following the genius AI assumption, ethical problems in relation to AI are often framed anthropomorphically, pushing the AI system into the role of the moral agent (see the discussion of 'machine ethics' in the Introduction). This means debating AI systems as a kind of agent that might one day populate our virtual or material world as new entities (think of troll bots on social media or self-driving cars traversing our urban landscapes), sharing our interpersonal space, interacting with us at eye level and potentially outperforming and replacing us in some skills. The ethical questions raised in these debates range from corner cases such as the trolley problem to the 'value alignment problem', which asks how AI systems could be built to 'align' in their interactive behaviour with the moral values that govern human behaviour (Peterson, 2019; Wong and Simon, 2020). In some cases, the question is even raised whether autonomous agents should be legally recognised as members of our societies that possess rights and responsibilities (Gunkel, 2018).

I will sidestep these popular AI ethics conundrums in the approach presented in this book. Without musing about whether animism, anthropomorphisation and other consequences of genius AI are philosophically right or wrong, my main argument is that they are dangerous and harmful as a presupposition in the ethics of AI debate. The spatial framing of AI as manifestation of intelligent behaviour localisable in autonomous agents does not reflect the status quo of AI technology at all. Measured by their relative urgency, issues such as the trolley problem and robot rights are comparatively irrelevant, and the framing of AI as contained in autonomous artefacts tends to distract ethical and political attention from the more serious and present implications of non-anthropomorphic AI. By this I mean applications of AI technology that millions of people worldwide are confronted with every minute – without facing embodied autonomous agents.

There are many currently widespread, already impactful applications of AI technology that do not fit in the genius AI scheme, and so easily go unnoticed in social and ethical debates, even though they have an enormous impact on many of us and would therefore be urgently worth discussing in terms of their ethical implications. Most AI technology today operates in data centres and on social media without facing us as virtual agents, robots or machines. Such technologies influence, for example, what news, advertisements or posts from our friends we see, what terms we are offered on banking or insurance, or whether we are invited to a job interview. We do not get to grasp the artefacts that make (or assist) these decisions as interactive agents. These AI systems are not interaction partners but rather *structural factors* in our digital communication (Mühlhoff, 2020b). They influence our access to information, resources and opportunities by being used to make predictions about our behaviour, our thoughts, feelings, illnesses, desires and needs, which enables these systems to deal with each of us in a 'personalised' way – and 'personalisation' is the popular euphemism for treating people differently.

In search of a more realistic and critical starting point for a definition of AI, the Organisation for Economic Co-operation and Development, for instance, states: 'An AI system is a machine-based system that is capable of influencing the environment by producing an output (predictions, recommendations or decisions) for a given set of objectives' (OECD. AI Policy Observatory, 2019). Using similar wording, the EU's AI Act states:

An AI system is a machine-based system designed to operate with varying levels of autonomy and that may exhibit adaptiveness after deployment and that, for explicit or implicit objectives, infers, from the input it receives, how to generate outputs such as predictions, content, recommendations, or decisions that can influence physical or virtual environments. (Madiega, 2024)

In terms of the phenomena I will address under the title of 'AI' in this book, I will roughly follow these definitions. As a consequence, I will focus on *data-driven AI systems* that are used to predict behaviour or unknown information about individuals, to allow the sorting of individuals into risk and scoring groups or to make personalised decisions in various areas of life. I will focus on these AI applications because they are the most common and, in terms of market capitalisation, the most important, and because they have enormous societal impacts in the form of social inequality, discrimination and manipulation.

Core examples of AI applications that I will refer to throughout the book include: search engines,[8] voice and facial analytics software,[9] rating and scoring algorithms for credit and insurance risks,[10] personalised advertising,[11] political targeting,[12] AI algorithms in human resource management and hiring processes,[13] AI algorithms in the security and police apparatus (predictive policing, border regimes, visa decisions),[14] in the education sector,[15] in youth protection[16] and in social benefits.[17] In all these areas, AI technology is routinely being used as a covert technology, not immediately visible and generally intangible to the end user.

To do justice to the current reality of AI technology, we need to look at non-anthropomorphic but high-impact AI systems like these examples. We must resist the lure of embodying, localising and futurising AI. Rather, an ethically fruitful understanding of AI must make visible the complex sociomaterial nature of contemporary AI. As we see in this book, contemporary AI relies on a specific form of relationship to humans. This relationship is built from power relations, relations of immersion, subordination and exploitation, relations that technologically shape human subjectivity and consciousness. As we will also see, there is a reciprocal codependence of humans and AI systems: not only does AI influence what we think, feel and do. Rather, our activities of thinking, feeling and acting themselves feed into the AI apparatuses of networked digital media. In the next section, I will flesh out what I mean by this bidirectional interconnectedness by means of a few examples and, building on this, introduce an understanding of AI systems as networked human–machine systems, or what I call 'Human-Aided AIs'.

Contemporary AI systems as hybrid computing networks

Perform a search with a search engine, such as Google Image Search, using a chosen keyword or phrase, such as 'pink elephant'. The outcomes of this query represent a manifestation of AI, as an AI system determines the images that meet the criteria of the search parameters. This exemplifies a typical AI system as it functions today, firmly grounded in the contemporary era, rather than residing in the realm of future developments. Moreover, it represents an AI system that has a substantial economic value, whose operator has secured a significant market share in search services. It further exemplifies an AI system that does not confront the user as a tangible, interactive entity; rather, it appears ethereal and spatially untraceable, operating in Google's data centres.

How does a computer system achieve the capability to generate images in response to a specific keyword? The challenge of machine image recognition, encompassing the transformation of visual content into textual descriptions, has long been a 'hard' problem in the field of machine intelligence research. To illuminate precisely how contemporary AI technology has cracked this nut and endowed image searches with their own purported 'intelligence', it is worth embarking on a brief historical journey to the early 2000s. The digital landscape of that bygone age bears little resemblance to today's. Smartphones and Facebook had not yet made their debut, and the dotcom bubble had recently burst, ushering in a wave of bankruptcies among internet-based business models and enterprises. Those companies that emerged as survivors at the end of the first internet era have since evolved into huge entities, including Amazon and Google, both of which saw substantial growth after 2000. Facebook also emerged on the scene in 2004.

At that time, Luis von Ahn, a doctoral student at Carnegie Mellon University in Pittsburgh, wrote a dissertation with the eye-catching title

Figure 1.2: Screenshot of the ESP Game (ca 2004, face obfuscated)

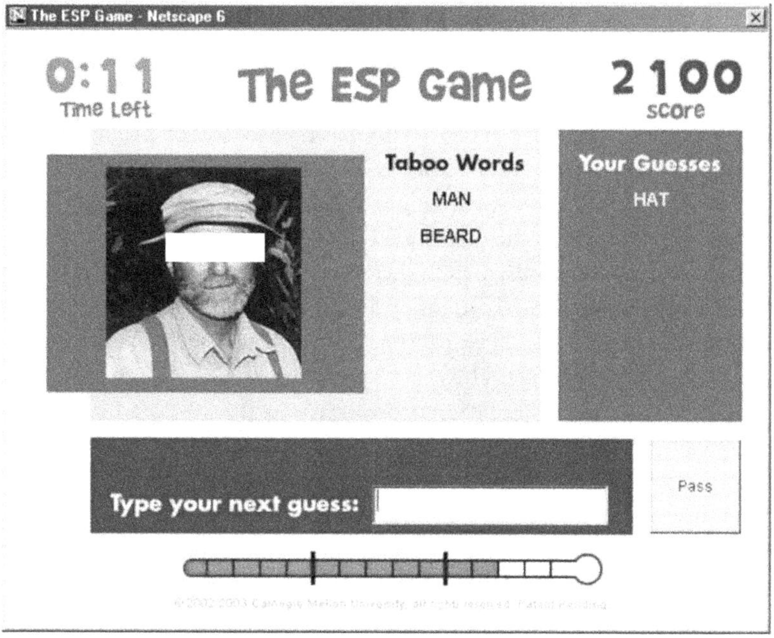

Source: ACM, Luis von Ahn and Laura Dabbish (2004)

© Luis von Ahn and Laura Dabbish. 2004. Labeling images with a computer game. In Proceedings of the SIGCHI Conference on Human Factors in Computing Systems (CHI '04). Association for Computing Machinery, New York, NY, USA, 319–326. https://doi.org/10.1145/985692.985733. Fig. 2.

'Human computation' (von Ahn, 2005; see also von Ahn, 2006a). So, what does human computation mean? Computation by humans, of humans, in a human way? In 2006, at the invitation of Google, von Ahn gave a 'Google tech talk' on his dissertation, summarising its main thrust by saying, 'We are [in this project] going to consider all of humanity as an extremely advanced, large-scale distributed processing unit that can solve large-scale problems that computers cannot yet solve' (von Ahn, 2006b: 8 min). So, in his dissertation project, von Ahn was doing nothing less than viewing the whole of humankind as a large-scale computational network.

As part of his project, von Ahn introduced an online game known as the 'ESP game' (see Figure 1.2) (von Ahn and Dabbish, 2004). This browser-based game was somewhat reminiscent of a fusion of the popular party games Taboo and Pictionary. In the ESP game, participants engage in tandem with another individual, their pairing being determined randomly by the server. However, direct communication is precluded between these participants, who generally reside in separate locations across the globe. The game encompasses multiple iterations of a consistent task: both the participant and their anonymous counterpart are presented with an image on their respective computer screens, and their shared objective entails identifying and inputting a keyword that aptly characterises said image. Points accrue to both participants when they successfully input identical keywords, with swifter responses yielding higher point values.

The game's design encourages participants to consider the most suitable keyword for a given image. What is the predominant visual aspect that comes to mind initially? What might the other person be most likely to think of when they see this picture? The game incorporates the provision of designated taboo words, which introduces an added layer of complexity by prohibiting their use. Consequently, players find themselves having to deliberate on alternative keywords that they deem to be the next-best terms to describe the image accurately.

Luis von Ahn and his colleagues brought out the ESP game in 2003. It quickly caught on, gaining almost 14,000 users after only four months. At first glance, this sounds like a successful business idea for an online computer game, but why did it feature in a computer science dissertation? To what end was the ESP game created? The real purpose of this game was to generate labels for the image content of a large database of images – in other words, to obtain, for any given image, an accurate list of keywords that in some way described or reflected that image's 'content'. In this way, the 14,000 'players' using the game in its first four months generated almost 1.3 million labels for around 290,000 images (von Ahn and Dabbish, 2004).

In the mid-2000s, there was no such thing as Google Image Search. Or rather, there was a service called Google Image Search, but it could not really access the pictorial content of the images to process a user's search query. If you searched for 'dog', Google Image Search would return image files on the internet that had the word 'dog' in the file name (for example, dog.gif, my-little-dog.jpg), or image files that were embedded in websites where the word 'dog' appeared in the surrounding text. The search algorithm used at that time did not have the capacity to analyse the content; it could only semantically evaluate the linguistic context of images.

In the days when computer systems had yet to achieve genuine image recognition capabilities, the ESP game was able to get its users to generate high-quality labels for an initial inventory of 290,000 images within a remarkably short time and without any associated costs. These labels constituted an ideal search index for this collection of 290,000 images, because the labels could then be deployed to deduce a connection between a sought-after keyword and the images linked to that keyword. Even the relative pertinence of images in response to a particular search term could be weighted based on the criterion of early occurrence of the keyword during ESP gameplay. And, of course, von Ahn did not need to pay users for their help in building this comprehensive index to his image repository, as it was an incidental (at least to the players) by-product of their engagement with the game. Not without sarcasm, von Ahn remarked in his presentation that some fans of the game might even have been willing to pay to play, given its popularity.

The ESP game, with its promise of potentially (and literally) game-changing advances in image recognition technology, was subsequently acquired by Google in 2006 and rebranded as Google Image Labeler. This interactive application was a useful tool in Google's broader mission to index its entire repository for Google Image Search, effectively serving as a training mechanism for their AI system dedicated to image retrieval. The Google Image Labeler boasted a substantial user base, facilitating the indexing of Google's ever-expanding database of images, which comprised 425 million items by 2004, within a mere six-month time frame. This achievement entailed the generation of approximately 27 new meaningful labels every second, all without the need for any financial remuneration to 'participants' in this labelling. Thus, the 'AI' underpinning this image search mechanism is fundamentally rooted in the collaborative efforts of a considerable number of individuals who, just by playing an online game, contributed to massively augmenting the image search software's AI capabilities.

This vignette is paradigmatic of the kind of AI systems that are the subject of the ethical debates and critiques to be found in this book. Opposing the anthropomorphising notion of genius AI, I assume that our relationship to AI systems is marked by immersion rather than mirroring, by inclusion rather than replacement, by the cybernetic logics of swarm intelligence rather than the mathematical paradigm of implementing the superhuman genius. Not only do the AI systems I scrutinise affect millions of us around the globe without even being faced as interactive agents, in truth they *rely* on us, both as collaborators and as data producers. I call this form of AI, as well the sociotechnical paradigm of analysing and criticising it, 'Human-Aided AI'.

2

Human–Aided AI

Human–Aided AI is a proposition for rethinking AI systems as sociotechnical systems.[1] As Katherine Hayles wrote in 2016: 'The most transformative technologies of the later twentieth century have been cognitive assemblages; the internet is a prime example' (Hayles, 2016: 34). Proceeding from the internet to the embedding of networked devices and services in every niche of life, these cognitive assemblages are now everywhere. While the public imaginaries of 'genius AI', fuelled by the industry-driven 'hype' of this technology (Chapter 1) are strategically hiding the sociotechnical nature of AI, Human–Aided AI is the direct proposition to rethink AI systems as sociotechnical systems. Human–Aided AI is a research paradigm for a critical philosophy and power-aware ethics of AI. In this perspective, the social, societal, ethical and political implications of data-driven AI take centre stage.

The Human–Aided AI approach illuminates the extent to which human 'collaboration' – through everyday use of the systems, involuntary data donations and new forms of digital labour – is an integral part of these systems. Although AI systems in part consist of technical components such as computer chips, networked devices, software and algorithms, the Human–Aided AI approach points to the essential role humans and human brains play in those systems as subcomponents. AI systems are thus conceived as hybrid human- and silicon-based computing networks. These hybrid networks, as a whole, accomplish the artificially intelligent skills that we mistakenly perceive as *purely technically generated* intelligent skills. So, when Google Image Search or DALL-E provide us with a menu of images to choose from, or when language processing tools like DeepL or ChatGPT generate text for us that we perceive as purely a product of machine computation, those systems are nevertheless relying heavily on human collaboration that often goes unacknowledged when we speak about *artificial* intelligence.

One of the main questions that arises from the Human–Aided AI paradigm is what actually drives humans around the world to become part of Human–Aided AI networks? The answer is: power. Human–Aided AI systems exploit different forms of power relations between humans and technological interfaces, as well as in the context of global economic and neocolonial power hierarchies, through which the various manifestations of our thinking, feeling and acting are captured with the aim of enabling and enhancing AI systems. These forms of power range from platform-based business models to the persuasive (and arguably addiction-encouraging) designs of digital interfaces or to the economic hierarchies exploited in online labour and the gig economy.

The Human-Aided AI paradigm rests on three main pillars or 'theses', roughly relating to three dimensions of power that together drive the embedding of humans in AI networks. The first thesis states that most contemporary achievements in AI would not be possible without business models of data extraction that lie at the core of digital services and products. The second thesis highlights how graphical user interfaces and user experience (UX) design play a central role in the construction of hybrid sociotechnical systems, because it is via interfaces – that is, intersections between networked data infrastructure and human bodies – that people and their cognitive performances are integrated into Human-Aided AI networks. The third thesis relates to the exploitation of local and global economic power differentials through new forms of 'click-work' and digital labour that force humans to integrate into computing networks as paid labourers.

The first thesis of Human-Aided AI

To start with an example, let's look again at Google, but this time at its familiar search engine. When using Google Search, you are also accessing an AI that generates a list of websites as search results in response to the keywords you have entered. These search results are supposed to be up to date and relevant. But how is an AI supposed to actually know and judge what is up to date and relevant right now? The answer is that Google Search receives this information from *us*, the users. Whenever you use the search engine, you become part of its AI system, which is a Human-Aided AI system. But why? And how does this process work?

When you click on one of the search results suggested by Google, not only do you open the page you selected, but something else happens, unnoticed, in the background. The link you click on is constructed in such a way that it does not lead directly to the desired target address but to a Google server (for more detail, see Mühlhoff, 2019b). This server then forwards you, lightning-fast, to the target address of the page you were originally seeking. This is how Google registers every single click on its search results: each time you click on a search result, you are led on a detour through one of Google's servers. In addition, the Google-generated link that is called up when you click on a search result contains a number of dynamic variables (or 'parameters', in more technical terms), which are transmitted from your browser to Google without you noticing. These parameters indicate, for example, whether it is the first search result that you call up or whether you have already looked at other results, or whether you have returned to the results page using the 'go back' button – which would be algorithmically interpreted as evidence of dissatisfaction with the previous results. Google can also register whether you switched tabs in the browser between different search results. It also records how far down you scrolled to find the search result you clicked on. And if you are logged into a Google account in another browser window at the same time (Gmail, YouTube or another Google service), Google even registers who you are via the parameters transmitted with the search result link – so that the search is consequently assigned to the personal search history of your account. After Google has registered these and many other parameters, you are then, in a few microseconds, redirected to the page you were originally intending to open, usually unaware of the roundabout path you have taken.

Tracking clicks is one way for Google to get feedback on which search results are the most relevant for certain groups of people. The millions of clicks thus collected every second form an enormously extensive dataset, which is constantly fed back as training data into Google's AI system that generates the search results. The click-tracking technology is thus a core piece of infrastructure that enables the Google Search AI to hone its capabilities – and all users are contributing to it with their clicks, constantly helping to calibrate and recalibrate the system so that it can remain up to date and always display relevant results, even in rapidly changing informational contexts (current trends, news events, political developments and so on).

All this makes Google Search a prime example of a Human-Aided AI system. It is important to note that human involvement is *continually* required in such AI systems. That is, there is no point in time when these systems have finished their training so that human involvement is no longer required. This also applies to the case of Google Image Search that was debated in Chapter 1. Here it would be easy to get the (wrong) idea that Google's image cache was eventually definitively labelled by means of the browser game, with human involvement no longer being necessary.[2] But this is far from the case, because the constantly evolving context of the real world, which is presumed to be accessed in this example by means of AI, also requires a constant recalibrating and retraining of the AI model.

The example of Google Search thus makes it clear that human support is generally not only an initial or intermediate stage in the development of data-driven AI systems, but rather an integral and permanent part of them. Put more bluntly, the computational operations required to produce search results in response to a search query take place partially on computer chips *and* partially in human brains. In other words, the AI system behind Google Search is an algorithm that orchestrates and combines both the computational operations on silicon-based computer chips and the 'read-out' of human reactions, behaviours and cognitive processes. This is the defining feature of Human-Aided AI systems: the computations that enact a system's intelligence are in part carried out in human brains. To go even further, *most* AI-based products and services today rely on the continued involvement of human brains. Users normally notice little of this at the front end, as the graphical or haptic human–machine interfaces we use in our day-to-day lives reveal little. In the background runs the code that processes the data we produce when we use the interface. The computational operations triggered by this code are in part executed in processor cores located in data centres, but, crucially, their execution relies in part on human input. This human input could be, for instance, the reaction of other users to a certain piece of content – for example, the users who performed a similar Google Search a few seconds or minutes earlier. So, to summarise, the first thesis of Human-Aided AI is:

Thesis 1: Most commercial AIs today rely on business models that integrate users into hybrid computing networks and exploit their cognitive capacities. Most contemporary AI systems are thus hybrid brain/silicon-based computer networks enabled in the context of contemporary media culture.

This may well be what Luis von Ahn meant by his ambitious announcement that he will 'consider all of humanity as an extremely advanced, large-scale distributed processing unit' (von Ahn, 2006b: 8 min).

Example: Facebook's DeepFace project

To dig a little deeper into the power strategies Human-Aided AI deploys to get users involved in its hybrid computation networks, we now consider another instructive example. Since the 2000s, the social networking platform Facebook has offered its users the function of tagging friends' faces in uploaded photos. In other words, that part of a photograph where a face is visible can be linked to the Facebook account of the photographed person (Figure 2.1). From today's perspective, this seems a completely natural and familiar practice. 'Tagging' faces in photos has become part of the firmly established forms of social interaction on many social media platforms and appears to offer itself as a well-loved feature for communicating with friends, enriching our photo albums with background information or putting a name to that face from the party last night.

But this 'natural' propensity to tag faces has not always been there. In 2007, one of the first years that networked online media such as Facebook and Flickr offered photo tagging, some UX designers were already observing that '[u]sers have mostly avoided annotating media such as photos – both in desktop and mobile environments – despite the many potential uses for annotations, including recall and retrieval' (Ames and Naaman, 2007: 971). In 2010, Facebook announced a project called Tag Suggestion and a company

Figure 2.1: Face tagging on social media. The user can select from an automatically generated shortlist of relevant users

Source: Adapted by Rainer Mühlhoff from Garry Knight (2012) 'Friends with Mobile Phones', licensed under CC-BY 2.0, available from: https://www.flickr.com/photos/garryknight/700 3178857/in/photostream/

goal of building a face recognition system to 'help' users tag people in images (Facebook, 2010; EPIC, 2011: point 48; Norval and Prasopoulou, 2017). The idea seemed to be that users would be more motivated to tag faces if all they had to do was select the appropriate person from a short list displayed as a pop-up window right next to the face. In 2012, Facebook acquired the Israeli start-up Face.com, which had been developing top-performing facial recognition algorithms since 2007 and had already been offering sophisticated face-tagging services via Facebook add-ons for several years. The two were a perfect match: Face.com brought its algorithmic know-how (with its cofounder Yaniv Taigman becoming a leading facial recognition researcher at Facebook after the acquisition); Facebook brought the data.

Clearly, large amounts of training data in the form of labelled images of faces are needed to develop AI for facial recognition. A popular social network like Facebook was in the best position to obtain such data for free if it could motivate its users to provide them voluntarily and as part of their social interactions on the platform. Thus, the more effective Facebook became at addressing the issue of users' reluctance to 'annotat[e] media such as photos' (Ames and Naaman, 2007: 971), the more efficiently the company was able to train its AI systems to recognise faces. At the same time, the reverse is also true: the better the AI is at facial recognition, the greater the capability of the Tag Suggestion function to encourage users to annotate faces, and therefore the greater the number of labelled face images that Facebook would be able to capture from its users.

This circular relationship between human engagement and service quality is typical of data-driven AI systems when we view them as Human-Aided AI systems. Such circular processes are designed to enable an AI system to 'pull itself up by its own bootstraps'. Andrew Ng, AI entrepreneur, former head of the Google Brain project and one-time chief AI scientist at Baidu, characterises the implementation of such data-harvesting cycles as the golden rule for bootstrapping any AI project: start with an initial version of the AI product that attracts some users. These users then generate data by using the product. The more data are gathered, the more the quality of the product improves – especially if this is a machine learning product that is constantly learning from the growing amount of data. The better the product, the more users it attracts, allowing the cycle to repeat itself over and over. In what seems like a nod to the idea of the 'vicious cycle', Ng calls such data-harvesting loops the 'virtuous cycle of AI', which is an integral part of AI business strategy because 'the scarce resource for any defensible AI project is training data' (Ng, 2017). Rather than providing a full-fledged service, AI companies launch incomplete or supplementary products with the primary goal of using these to collect data. One way of doing this has been dubbed 'pseudo-AI', referring to practices that launch a product that is only putatively AI, but that in reality has human workers performing the AI task at the back end, while the user interface maintains the illusion of 'smart' technology (Solon, 2018b). By recording the actions performed by click-workers, an initial set of training data can be collected to eventually train a machine learning model.

Facebook's face recognition system serves as a model case for this strategy, as bootstrapping this AI project initially required overcoming two interrelated hurdles: acquiring enough data to train the machine learning model and overcoming user unreliability in generating these training data. Facebook's catalyst for setting in motion a reciprocal process to resolve these two conditions was to integrate face tagging into users' everyday social interactions on its primary platform. The platform framed face tagging in such a way that users did not simply tag faces to add details to their own photo albums. Instead, face tagging was *itself* turned into a form of social (inter)action. The tagged person was notified and could share the event on their timeline, which attracted the attention of other Facebook friends. Being tagged in a photo thus became a key component of the specific logics of recognition, belonging and social communication created by Facebook. As a catalyst for initiating the 'virtuous cycle' for training an AI for face recognition, Facebook thus exploited its users' social motivations.

Once this self-reinforcing cycle of covert user engagement had started to take off, Facebook succeeded in creating a top-performing AI system for facial recognition, finally announcing it in 2014 under the name DeepFace (Chowdhry, 2014; Taigman et al, 2014; Bohannon, 2015). For many years, Facebook downplayed and even concealed the fact that the platform was exploiting users' photographs and labels without their consent to extract biometric information and to train an AI system on these data (Norval and Prasopoulou, 2017). Already in 2011 this led to the filing of a complaint by the Electronic Privacy Information Center and other US civil society organisations to the Federal Trade Commission in Washington, DC (EPIC, 2011). Facebook responded to these objections by reinterpreting its project in the light of growing public awareness of privacy issues, stating that the intention behind its face recognition technology was not to invade but to protect users' privacy: once the face recognition AI identifies you in an uploaded image, 'you will get an alert from Facebook telling you that you appear in the picture [...]. You can then choose to blur out your face from the picture to protect your privacy', explained Yann LeCun, a director of Facebook's AI research division, according to Bohannon (2015). This strategy of embracing and appropriating criticism and objections is typical of platform companies. It diverts public attention from substantive issues of privacy and data protection, namely that *the platform* has the capacity to identify faces.

Facebook put LeCun's announcement into action and in 2017 took the DeepFace project to another level. The face recognition model was now being used to automatically label faces in cases when a person in a photo had *not* been tagged by a user. In these cases, a user whose photo had appeared on someone else's post would receive a notification that someone had 'added a photo that [they] might be in'.[3] This 'tag confirmation dialogue' presents the user with the choice to agree to be identified in the picture, to be 'ignored' (that is, to not be tagged even though the identification is correct) or to state that they do not appear in the picture in question. The announcement of this feature suggested DeepFace had been created to give users more control over their privacy. After all, 2017 and 2018 saw the aftermath of the Facebook/Cambridge Analytica scandal (see Chapter 5), which had put Facebook under intense pressure on privacy issues. Public sensitivity to questions of manipulation based on users' data traces on social media was widespread at the time. The advertising strategy for the facial recognition feature shows how powerful platform companies were attempting to exploit and reinterpret this growing privacy awareness among their many users for corporate gain.

In the case studied here, Facebook even managed to exploit precisely this sensitivity to privacy concerns to trick users into producing *extra* training data for the AI system in question. The real purpose of the tag confirmation dialogue box, much like the Tag Suggestions function, was presumably to generate *verification data* for the AI system – that is, a constant stream of feedback data from humans to verify and, if necessary, retrain the face recognition model. The Tag Suggestions function presents the user with a short list of possible identities that match the face; this is ideal when the system is still at an early stage in which it cannot yet single out *one* face with a wide confidence margin. The tag confirmation dialogue box is geared towards the 'adult' stages, where the system autonomously generates tags but still needs feedback loops to remain reliable in a world of constantly emerging new faces. The first two choices in the confirmation dialogue – whether the user wants to be tagged or remain incognito – make no difference to the AI system, as they both confirm that the image in the photo is that of the person themselves and that automatic detection was therefore successful. Relevant for the AI system are the first two options as opposed to the third – the possibility of erroneous recognition. The information about correct vs incorrect recognition, once confirmed by the user, can be fed back into the system as training data to enhance the efficacy of the AI model.[4]

The second thesis of Human-Aided AI

The historical example of Facebook's DeepFace system is further confirmation of the first thesis of Human-Aided AI, as it shows how part of the intelligence performance of Facebook's face recognition system has always come from users. However, the example also shows something more: the way users are harnessed to become unwitting contributors to this AI system proceeded through specific elements of HCI design. Both the Tag Suggestions pop-up (as of 2011) and the tag confirmation dialogue function (as of 2017) are examples of UX design that facilitates Human-Aided AI. The design of these interfaces aims to engage users without their knowledge to produce training and verification data for this particular AI system. Indeed, a significant, functional part of Facebook's DeepFace system was implemented through expertise in the fields of HCI and UX design. To put it even more pointedly: *a significant part of the 'intelligence' of Human-Aided AI systems lies in the design of feedback loops.* We see that AI systems are products not only of programming savvy, but also of craftsmanship in the field of interaction design. This holds for most contemporary examples of data-driven AI, as almost every challenge in the field of AI seems now to be accompanied by the question of how to push users engaging with addictive interfaces to produce data.

This illustrates the second thesis of the Human-Aided AI paradigm, which I call the *correspondence principle*:

Thesis 2: There is a tendency to translate machine learning problems into human–machine interaction design problems. That is, problems in machine learning (building a particular model) are solved by finding a solution to a corresponding problem at the level of interface design (getting users to create the relevant training data).

The story about Facebook's face recognition program also highlights an approach to the generation of user-sourced training data that is very similar to the ESP game (see Chapter 1), but without the element of 'gamification'. The ESP game was explicitly played by users as a game; they were also informed that the data obtained through the game would be used to train image recognition systems. In the case of Facebook's photo tagging, users were *not* told that their tagged images would also be used to train a face recognition AI; neither were their play-oriented impulses exploited, but rather their motivations for social interaction on the platform were, including their psychological needs for recognition and belonging in social relations. By channelling these social impulses, Facebook has *created* a social sphere in which tagging faces in photos is part of everyday interaction. The initial perception of users who found it too cumbersome to annotate images online shows that it is not at all self-evident or 'natural' that users sit in front of their screens and mark faces on photos. Rather, this is a purposefully constructed landscape of social media interaction – the purpose for the company being to collect AI training and verification data through users' unwitting cooperation (Mühlhoff, 2020c).

Click-work

So far, I have described strategies of capture that harvest AI training data as a by-product of everyday and voluntary online activities. In principle, this mode of extraction exploits users of digital services around the world and from all strata of societies – including users in the so-called Global North and South,[5] in affluent and less affluent (or privileged and unprivileged) societies and in strata within those societies. While analysing and critiquing the value extraction from user data that is a key aspect of the Human-Aided AI approach, contemporary AI also relies on other forms of data capture and exploitation that are based more on domination and economic disparities. Since the early days of 'human computation', approaches like gamification and crowdsourcing have been complemented by paid microtasking and click-work – that is, by data acquisition in a new kind of alternative labour market. In making use of click-work, tech companies from the Global North leverage economic and political hierarchies, as well as international differences in wage levels and employment legislation, to profit from workers labouring under conditions far below the standards of the countries where these tech companies are headquartered (Roberts, 2019; see also Haidar and Keune, 2021).

Click-work as a term refers to the idea of having humans perform minor, repetitive on-screen tasks, such as labelling data, annotating text and images or transcribing audio, from their computers or mobile phones for a small per-item compensation.[6] A pioneer of click-work has been Amazon with its Mechanical Turk (MTurk) platform. This service was launched in 2005, around the same time that Luis von Ahn and his team were developing their idea of extracting data-related work for free through gamification. MTurk was originally developed for Amazon's own purposes, as a computing infrastructure designed to outsource a number of repetitive tasks related to maintaining its extensive product catalogue, such as updating product information and identifying duplicate goods. In the jargon of the platform, small tasks that can be processed by humans in a few seconds for a few cents are called Human Intelligence Tasks (HITs). On MTurk, a global community of casual workers is always available to process HITs submitted by large companies or research institutions through an application programming interface (API). This community of workers is mostly located in the Global South or in economically precarious strata of societies in the Global North (Roberts, 2019).

The name 'Mechanical Turk' contains a reference to a historical curiosity. The Mechanical Turk was a fraudulent chess-playing machine designed and built in 1769 by the Austrian Baron Wolfgang von Kempelen at the court of Maria Theresa (Levitt, 2000). The stunning and famous automaton, which von Kempelen demonstrated in many public shows and which toured Europe and the Americas for several decades, created the illusion that it was able to play an advanced game of chess, when in fact a human chess player was concealed inside. The Mechanical Turk consisted of a cabinet-like base within which the human chess player could hide; on top of the cabinet was a chess board, behind which sat a life-sized model of a person. The puppet was dressed in a turban and traditional Ottoman robes, which suggests the popular name 'Mechanical Turk'. It is important to know that in the original German, the term *Türke*, while commonly referring to a person of Turkish origin, also has a pejorative association with the derived verb *türken*, meaning 'to fake [something]'.

The way Amazon took up the historical reference to the fake chess automaton has a whiff of cynicism about it, not only in its reproduction of a derogatory reference to Turkish people. With the MTurk platform assembling and providing access to a global human microtasking workforce via an API, the service fully aligns with the idea of 'human computation', because it effectively provides a software interface designed to connect with human brain power. In 2014, the computer scientist Jaron Lanier wrote:

[Amazon's Mechanical Turk] is a way to easily outsource—to real humans—those cloud-based tasks that algorithms still can't do, but in a framework that allows you to think of the people as software components. The interface doesn't hide the existence of the people, but it still does try to create a sense of magic, as if you can just pluck results out of the cloud at an incredibly low cost. (Lanier, 2014: 170)

Through the MTurk API, the processing of HITs on a 'human processor' can be seamlessly integrated into classical programming code – which is then executed on a hybrid architecture combining silicon chips *and* human brains. The inherent entitlement and cynicism underlying this form of neocolonial appropriation of labour is reflected by Lanier in his continuation of the passage previously quoted:

> The service is much loved and celebrated [...]. My techie friends sometimes suggest to me in all seriousness that writing books is hard work and I should turn to the Mechanical Turk to lower my workload. Somewhere out there must await literate souls willing to ghostwrite for pennies an hour. (Lanier, 2014: 170)

The MTurk service is regularly used for data annotation and labelling tasks by commercial actors and public research institutions alike. In fact, the availability of click-work platforms like Amazon's Mechanical Turk is as central to the emergence of contemporary data-driven AI as are the ideas of gamification and implicit data capture based on UX design.

Click-work is just one of the more egregious manifestations of how embedding humans in hybrid computing systems leverages economic power differentials on a global scale. It is important also to bear in mind that the kind of power involved in click-work is different from the 'soft' power of nudging and UX design that tricks users at the front end (privileged people included) into lending their brain power to hybrid computing networks. Access to human workers via an API largely obscures to MTurk's users those economic power differentials and societal inequalities that enable the availability of click-workers at the back end, as well as the social consequences of this digitalised exploitation of human and cognitive resources.

Commercial content moderation and the social costs of click-work

The social consequences of click-work are particularly evident in the example of commercial content moderation (CCM). CCM refers to the outsourcing of moderation tasks by social media platforms to service companies that rely on click-work for manual reviews of user-generated content such as posts, images and other uploaded material (Roberts, 2016a). The army of CCM workers deployed by a social media platform such as Facebook has been conservatively estimated at several tens of thousands of people across the world (Roberts, 2019; Steiger et al, 2021). These workers review user-generated content item by item, day by day. Most platforms do not send all user-uploaded posts or images for this type of manual review, as this would be very expensive. Typically, user-generated content is posted online immediately, and only once other users have flagged it as inappropriate is it then forwarded to CCM staff. This filtering mechanism, where only flagged content is forwarded for review, means that CCM workers do not see the full range of uploaded material, but rather a preselected list. As Sarah Roberts points out:

> [this material] often focuses on content that is highly sexual or pornographic, depicts the abuse of adults, the abuse of children (physical and/or sexual), the abuse and torture of animals, content coming from war zones and other areas besieged by violent conflict, and any material that is designed to be shocking, prurient or offensive by nature. (Roberts, 2016b)

Investigations by journalists and researchers point to psychological damage, such as post-traumatic stress disorder, resulting from this work (Chen, 2014; Krause and Grassegger, 2016; Roberts, 2016b; Roberts, 2019; Steiger et al, 2021). This is a social externality of click-work that contributes to the expropriatory working conditions in the gig economy and tends not to feature in the balance sheets of multinational companies that use these types of moderating services.

Even though platforms like Facebook claim that they are increasingly using AI for moderation tasks, the involvement of human labour remains central to content moderation systems.[7] Imagine that a given platform employs some sort of machine learning–based classifier to make moderation decisions. If the classifier is very confident about its assessment of an item of user-generated content, that result will directly be used to remove that content. There will, however, be many uncertain cases (that is, inconclusive output from the classifier) or cases in which the content, after passing through the automated system, is flagged again by users – either by other users as 'inappropriate' content, or by the poster or creator of the item complaining that their content has been unjustifiably deleted. In this case, where the automatic classification has allegedly failed or is inconclusive, the platform automatically sends the item for a click-worker to manually review, not only to make the final decision but also to gather training data that help to refine the classifier's AI model. Thus, the human click-worker's decision, triggered by the intervention of human users, serves as training data for the machine learning model used in the first step of automatic classification. In this way, different types of human intelligence capacities along with silicon processor–based machine learning models are incorporated into the content moderation system, making it a Human-Aided AI system.

We see that click-work is central to the development of modern data-driven AI technology and has been for the past 20 years. Click-work, often conducted via Amazon's MTurk program, is a source of cheap and fast data-labelling and annotation labour that is widely used in many AI development and research contexts, including those that are not user-centric. Research departments at universities in the EU and around the world routinely use click-work services, as does industry.[8] Recently, for example, it was revealed that OpenAI contracted Kenyan click-workers through a company called Sama to complete data annotation tasks for the large language model GPT-3, which is a foundation for their chatbot ChatGPT (Perrigo, 2023). This is unsurprising as most data-based AI systems rely on input data generated by humans.

The third thesis of Human-Aided AI

It is at the core of the Human-Aided AI approach to conceive of the relation between the capturer (in my examples mostly tech companies, but it could also be a state actor) and the user as a relation of *power*. On the spectrum of techniques of data capture (see also Mühlhoff, 2020c), click-work shows particularly blatantly how economic power differences, shaped by precarious working conditions and global economic inequalities, can be transformed into computing power. One can view a click-work platform like MTurk as a machine that converts economic power into computing power to run what at the front end is perceived as 'artificial' intelligence.[9] Diving further into this relationship between capturer and captured, it may be helpful to draw upon a notion by Nick Couldry and Ulises Ali Mejias, who state that the extraction of human cognitive capacities for AI 'is operationalized via data relations, ways of interacting with each other and with the world facilitated by digital tools' (Couldry and Mejias, 2019: xiii). Data relations are strategically established – in the form of new digital apps and services that users *want* to use, or in the form of digital interfaces that mediate microtasking and click-work. What all these strategies have in common, though, is that they operate *across* digital interfaces – interfaces by means of which a person is linked with an AI apparatus. These data relations across interfaces are thus driven by power mechanisms, and these power mechanisms can take vastly different forms, ranging from the soft power of nudge-based and addictive UX design (see Chapter 3) to a more domination-based power that exploits precarious working conditions and global economic inequalities. This power aspect of data capture is summarised in the third thesis of Human-Aided AI:

Thesis 3: Human-Aided AI systems are apparatuses that convert power differentials into computing power. Human-Aided AI apparatuses harness *various* forms of local and global power relations by means of interfaces that turn people into subcomponents of large-scale, distributed computing networks, thus leveraging human cognitive capacities for use in AI systems.

AI extractivism and digital colonialism

Human-Aided AI, insofar as the phenomenon characterises an essentially capitalist exploitation of human abilities and resources, adds a specific dimension to the current discourse surrounding 'AI extractivism' (Crawford, 2021). AI extractivism, in turn, builds on the more general debate regarding extractivism in post-Marxist critiques of capitalism. In its most commonly used sense (see also Gago and Mezzadra, 2017; Mezzadra and Neilson, 2017), extractivism refers to the workings of global capitalist structures that operate through the extraction of 'huge volumes of natural resources, which are not at all or only very partially processed and are mainly for export according to the demand of central countries' (Acosta, 2015).[10] In view of recent developments in capitalism, it has been suggested by Sandro Mezzadra and Todd Neilson (Mezzadra and Neilson, 2017) that the concept of extractivism should be extended from its traditional focus on the extraction of raw materials (for example, through mining or agricultural industry) to the multiple 'new frontiers' opening up in contemporary capitalism, particularly including digital capitalism, platform companies and novel kinds of internet labour (see also Mezzadra and Neilson, 2017: 9ff). This increasing relevance of the appropriation and extraction of data rather than only raw materials is one of the 'current processes of capitalist transition and upheaval' that call for a broader interpretation of extraction that encompasses 'new forms of exploitation that directly target social cooperation' (Mezzadra and Neilson, 2017: 13).

It is worth linking this analysis with the concept of 'digital colonialism' as a way of contextualising the global architecture and colonial legacy of data extractivism against the backdrop of rapidly evolving AI systems. In analyses of digital colonialism, scholars have pointed out that Big Tech practices of data extraction must be seen in continuity with the exploitation of global economic power relations that have been shaped by centuries of colonialism. Danielle Coleman, for instance, argues that African countries are exposed to particular costs and risks relating to data extrication by Western tech companies due to their 'limited infrastructure, limited data protection laws, and limited competition' (Coleman, 2019: 418).[11] In an allusion to the 19th-century 'Scramble for Africa' – the invasion, annexation, division and colonisation of most of Africa by Western European powers, culminating in the Berlin Conference of 1884/85 – Coleman points out the contemporary 'scramble' among US American tech giants like Meta and Google for the data of those living on the African continent. Through building or buying up critical communication infrastructure such as mobile data connectivity, these companies seek to extract usage data from African populations. 'With over 1.25 billion people living in Africa, this market represents [from the perspective of tech companies] a treasure trove of data, much of which is as yet untapped' (Coleman, 2019: 425).

As an example, Meta Platforms' (formerly Facebook's) Free Basic service has partnered with mobile internet providers in 32 African countries to provide users with free access to the Facebook platform in an environment where general data plans, providing neutral access to the whole internet, are expensive and generally unaffordable to most people.[12] This strategy has effectively turned Facebook into 'the internet' from the perspective of most users in those countries, enabling the platform to collect data on all communications from social to economic and political (Coleman, 2019). Because of the lack of competitors to Meta's services and the price lock-in effect, users cannot effectively escape Meta's data-capture strategies. Thus, while many internet users in the Global North encounter the subtle UX design strategies mentioned in the previous subsection that nudge them into providing their cognitive resources, users in the Global South, as well as those occupying more precarious positions in societies of the Global North, might be facing a much more rigid power structure of coercive extraction with respect to their data.

Adopting the notion of data colonialism in a critique of data capture and extractivism on the part of AI companies adds a key aspect to our analysis: that of the *naturalisation* of data as a resource that is just there, ready to be harvested and appropriated by whoever has the resources to reap and winnow it. As Couldry and Mejias argue:

> [The colonial mode of] [a]ppropriation frames resources as *naturally* occurring, free for the taking. Let us recall that historical colonialism dispossessed indigenous people of their land *before* they could conceive of it as private wealth in the way that the colonizers conceived of it as private wealth. [...] By the time indigenous people understood the concept of private property, it was too late: their land had acquired new value, this value had been stolen from them, and they themselves had been overpowered and enslaved by the new system. In a similar vein, our social lives are not material 'wealth' to us, like money and property are. But once they are datafied, they cease to be just life and become a source of wealth. (Couldry and Mejias, 2019: 88)

Data capture, both when it relies on the 'soft' power of UX design and when it is effected through domination-based strategies of colonising digital infrastructures in the legacy of historical colonialism, follows the idea that for-profit datafication of people's personal lives, as well as of their political, economic and social communications, putatively means just gathering an otherwise unused, unclaimed resource. This attitude of *entitlement* was already visible in the quotation from Jaron Lanier on Amazon's Mechanical Turk mentioned earlier, where Lanier described how his 'techie friends' would make uninhibited use of this service to ease their workload, even when it came to ghostwriting a book (see Lanier, 2014: 170), which could go as far as expropriating the intellectual property of click-workers. In the context of colonising infrastructure projects such as Free Basics, this neocolonialist mindset of hegemonic entitlement often comes paired with patronising narratives of altruism and benevolence that promote tech companies' supposedly 'free' offerings as a service not just to society, but to humankind as a whole.

As a key takeaway, the concept of data colonialism thus highlights that tech companies' strategies of domination and exploitation rely on the *naturalisation* of personal and social data as unclaimed resources. The way Couldry and Mejias introduce the term 'data colonialism', however, goes one step further in claiming that colonialism reaches a qualitatively new dimension in the digital context. In their framing, the term extends beyond any specific geographical context that might be implied by its link to traditional colonialism.[13] As Couldry and Mejias argue:

> In this neocolonial scheme, the colony is not a geographic location but an 'enhanced reality' in which we conduct our social interactions under conditions of continuous data extraction. The resources that are being colonized are the associations, norms, codes, knowledge, and meanings that help us maintain social connections, the human and material processes that constitute economic activity, and the space of the subject from which we face the social world. (Couldry and Mejias, 2019: 85)

Crucially, the colonisation of these resources occurs worldwide, although hugely diverse power mechanisms are employed in this process, ranging from the soft power of UX design to the leveraging and reinforcing of economic dependencies and global disparities in the legacy of historical colonialism. In a similar vein, Jim Thatcher et al prefer to use the term 'digital colonialism' to depict an economic structure of appropriation that is not immediately linked to specific geographical zones. They argue that the 'metaphor of data colonialism' allows us to discern the true patterns of domination and exploitation that usually lie hidden in the common trope of 'big data—and the "digital" in general—as new frontiers to be explored, expanded, and conquered' (Thatcher et al, 2016: 992). Viewed through this lens, the traditional project of colonisation has undergone a worldwide transformation and expansion over the past two decades, with the social lives of people everywhere becoming the next 'frontier'. Digital colonialism adds new virtual territories and new digital media–based relationships of domination and extraction to those global relationships of domination that already exist as part of the legacy of historical colonialism.

How do all these concepts relate to Human-Aided AI? First, as a particular offspring of extractivism in general, AI extractivism comes with the proposition that AI fundamentally relies on multiple dimensions and relations of extraction, both local and global, including the extraction of raw materials (needed to build computing hardware and end-user devices), energy and labour, but also of data originating from multiple populations and parts of the world. Digital colonialism adds to this analysis the key aspect of *naturalisation* of data as a resource that is being appropriated by tech companies, thus robbing individuals and societies of this resource before they even conceive of it as a potential source of their own wealth and prosperity. This strategy of naturalisation comes with the attitude of *entitlement* potentially reflected in the personal conduct of many tech engineers and entrepreneurs worldwide, and with the ideology of big data as a frontier, leading to (usually) Western 'pioneers' seeing themselves as being entitled to harvest all the putatively unclaimed resources on that frontier, much as was done in the days of traditional colonialism. Within this broad picture, the Human-Aided AI paradigm zooms in on the data-capture aspect of AI more closely, detailing the multiple forms of power relations and subjectivities involved in data capture. Human-Aided AI not only marks AI capitalism as an extractive and colonialist project, it also characterises the technology of AI as less purely 'electronic' than we might think. While the concepts of appropriation, expropriation and extraction highlight the (forced) transfer, displacement or stealing of resources, taking them from the users or workers and transferring them to multinational tech companies, Human-Aided AI expands this perspective by making a statement about the essence of AI, emphasising the fundamental incorporation of users and workers into AI systems as an operational necessity. Human-Aided AI thus invites a sociotechnical perspective on what AI really is, claiming that AI apparatuses are not only made from wires, semiconductors and storage devices, but are sociotechnically embedded, collaborative intelligence networks, deeply entangled with capitalist exploitation and new forms of profit, and as such fuelled by the collaboration and labour of humans worldwide.

Digital Counter-Enlightenment and the Power of Design

We have seen how the culture of networked digital media that has emerged over the past 20 years has produced a networked infrastructure and a variety of cultural techniques of interface design for the centralised collection of data from all sectors of society. This media culture has established digital services, use cases, interface designs, protocols and techniques to capture data streams coming from users and click-workers, while the dominant perceptions and narratives of our digital culture mostly fail to acknowledge data capture as representing the back-end purpose of many networked media services. The entanglement of media culture and data aggregation emerging since the 2000s has enabled, with a decade's delay, the kind of data-driven, machine learning–based AI we see flourishing today. It is now possible to examine the genealogy of AI, which is a story of power and extraction, as a media history.

In this chapter I will look more closely at the knowledge practices that constitute the power of interface design. We will see that human–computer interaction (HCI) design is itself a data-driven epistemic regime. As such it inscribes ethical and political qualities in networked media artefacts.[1] Using the term 'digital counter-enlightenment', I will point to a significant aspect of the politics of subjectification of mainstream user interfaces: while the relevance and pervasiveness of digital devices has grown in all areas of life during the past 20 years, instrumental rationality when it comes to using these devices, as well as the collective level of understanding of the inner workings of digital devices, have not grown at a commensurate rate (or have perhaps even decreased) over the last 20 years. I will argue that this is not only due to the technologies becoming more complex, but is also a result of infantilising design practices on the part of the industry. Digital counter-enlightenment manifests itself in a particular relationship between technical artefacts and their users, which is itself a consequence of how these technical devices have been designed. I will argue that digital interfaces today tend to be designed so that the autonomous and instrumental use of digital machines is systematically impeded or prevented. This gives rise to a form of subjectivity of users, which can then be exploited as part of the power strategy that is at work in digital media apparatuses.

AI in the history of networked media

In the 1990s, there was much talk of so-called ubiquitous computing ('ubicomp' for short). The term was coined by computer scientist and Silicon Valley visionary Mark Weiser to refer to his vision of computers, electronic sensors and computational information processing becoming 'so ubiquitous that no one will notice their presence' any more (Weiser, 1991). Computers and digital data processing would 'weave themselves into the fabric of everyday life until they [were] indistinguishable from it' (Weiser, 1991). Since the 2000s at the latest, Weiser's vision has been borne out, with computers entering every niche of life, every pocket, every refrigerator, every private, economic and public communication. Microcomputers, networked sensors, the Internet of Things and digital media infrastructure are spreading throughout the world, simultaneously penetrating every part of our lives.

However, while the development we have observed since the 2000s confirms Weiser's vision about ubicomp, it also features a trend that goes beyond Weiser's prediction: what is spreading most efficiently is *networked* digital media *interfaces*. This is not just about the gadgets in our pockets and on our wrists, or 'smart' refrigerators and lights. These gadgets are above all else *networked* devices and appliances that build interfaces between a ubiquitous data network and non-digital entities like human users and washing machines. It turns out that those digital devices and services that have been most prevalent and commercially successful (that is, most financially viable for their creators to produce) since the 2000s are precisely those that feature both a network and an interface aspect that enables a networked extraction of data from nearly everywhere on earth. The smart refrigerator that is smart only to the user cannot be monetised as effectively as the fridge that is really, from a corporate perspective, an interface with the user's kitchen, allowing companies to aggregate data from tens of millions of households. The smartphone laden with apps becomes an expansive business model only once it has been designed and produced as a network-connected interface connecting millions to the cloud.

Hence, what Weiser did not foresee when he confidently predicted that computers would 'vanish into the background' (Weiser, 1991: 3) as they sneaked into every nook and cranny of our lives is the converse development that was happening at the same time: *users were vanishing into computing systems.* Users were becoming ever more tightly integrated into large-scale distributed

Table 3.1: Historical landmarks in the emergence of networked media services for training data aggregation over the past 20 years

Year	Landmark
~1991	Ubiquitous computing
1997	Google Search
2003	ESP game
~2004	Web 2.0
2005	Amazon's Mechanical Turk
	Google Analytics
2006	Facebook
	Twitter
~2010	Smartphones
~2016	Deep learning hype

computational networks. For every small computational task that is hard to solve in traditional computational terms, there is now a way (a product, a game, a gadget) for humans to be incorporated into a computing system to address that task or problem. The various strategies of data capture are thus the condition of *ubicomp*'s economic reality in the 21st century. This reality can be seen in milestones such as the development of Google Search in the late 1990s; the Web 2.0 paradigm starting around 2004 (O'Reilly, 2005); the emergence of crowdsourcing and click-work in various forms, with their effects evident in the ESP game or in Amazon's Mechanical Turk, for example, around the same time; the establishing of tracking infrastructures using cookies or Google Analytics software; or the boom in social media and connected phones from the mid- to late 2000s (see Table 3.1).

Thus, in the 2000s we were witnesses to a technological development that has produced a ubiquitous and finely differentiated infrastructure of interconnected interfaces that are designed to integrate *everything* and *everyone* into data networks. This trend made 'deep learning' and other intensively data-driven AI techniques commercially successful roughly ten years after the spread of the datafication infrastructures that were its prerequisite (Mühlhoff, 2020c; Whittaker, 2021). It is therefore no coincidence that, although the conceptual and algorithmic groundwork for artificial neural networks and deep learning had been laid in the 1980s (see also Goodfellow et al, 2016: 13–21), it was only in the 2010s that the fanfare around 'machine learning' grew louder. It is only as a result of the sociotechnical transformation often referred to as 'digitalisation' that we now find ourselves in a situation where the current wave of AI favours data-based approaches rather than the symbolic, logico-deductive, 'Good Old-Fashioned AI (GOFAI)' paradigms that took centre stage throughout much of the second half of the 20th century (see Haugeland, 1985: 112).

Illuminating the genesis of contemporary AI in the context of digital media culture will make it evident that the critical paradigm of Human-Aided AI must always include historical or, more accurately, *genealogical* work.[2] Contemporary AI relies on an epoch-making shift in digital culture – that is, in the way people around the world enact their lives, social relations and subjectivities in and through digital communications media. We all are drivers of this transformation inasmuch as we endorse networked digital media services through our day-by-day use of them. With a view to its sociotechnical and power-relational preconditions, the historical and genealogical dimension of Human-Aided AI sheds light on the question of how the state of the AI art emerged as a (by-)product of media design. This makes a critical history of interface and HCI design hugely relevant to any ethics and critique of AI. As I argue in this book, to the extent that this is a critical project, an ethics of AI *must* be rooted in a media-genealogical perspective that foregrounds the operations of power in HCI design.

In contemporary media culture, the design of digital artefacts is inseparable from questions of expertise, professionalism, strategic intent and empirical methodology. Nothing is left to chance because design *is* political. In the following sections I will explain the politics of design in the culture of digital counter-enlightenment using various examples: nudging presents implicit and subtle ways of influencing users; tracking and web analytics make up a full-fledged infrastructure to covertly measure and predict user behaviour for the purpose of optimising interaction design; user experience (UX) design and what I call the trend of 'sealed surfaces' bear witness to a broader cultural decline in our instrumental relationships with technical devices.

Interface nudges and digital choice architectures

In May 2018, Google announced a new version of its smartphone operating system Android, which for the first time included a set of features to promote 'digital well-being'. For instance, a dashboard was introduced in the settings menu that shows users the time they have spent during the day on each installed app. In this way, the user would allegedly achieve an increased 'awareness' of their usage and gain insights into the risks of engaging in possibly addictive behaviour. At the same time, this feature represents a doorway for Google not only to measure usage and consumer behaviour in minute detail, but to be openly seen to be doing so for the supposed benefit of the user.

By the end of the 2010s, public debate was beginning to criticise smartphones as attention-demanding devices that were keeping people from sleeping, working and socialising offline. According to a report from *The Guardian*, the call for digital well-being came at a time of increasingly intense debate about the 'habit-forming design practices that encourage people to spend more time on their devices, such as infinite scrolling, notifications and other behavioural "nudges"' (Solon, 2018a). With the switch to allegedly more 'empowering' interface design, Google appropriated what had originated, with slogans like 'time well spent' or 'humane technology', as a backlash among some professionals working for Silicon Valley–based firms.[3] The end of this decade made use of usage tracking as a kind of mirror that reflected to users the wrinkles and contours of their own digital selves and gently took care of the user by means of subtle reminders and nudges that appeared everywhere. For example, the YouTube app featured the prompt 'Time to take a break?' if the user had been watching videos for longer than a predefined amount of time. And the Android screen switched to greyscale at a set time at night, with the phone automatically switching to do-not-disturb mode to lull the user to sleep.

All these examples are classic nudges in the sense of the term as it is often used in behavioural economics. Nudges are small, non-binding interventions that attempt to influence the behaviour of users 'for their own good'. The term 'nudge' as a behavioural intervention was coined by the influential work of the same name by Richard Thaler and Cass Sunstein (Thaler and Sunstein, 2008) as a comprehensive programme to transfer insights from psychology and behavioural economics to the field of public policy. Nudging has been described by some of its advocates as 'libertarian paternalism'. The idea is 'to steer people's choices in directions that will improve their lives'. Protagonists characterise this paternalism as 'libertarian' because of its 'weak, soft and nonintrusive' mode of intervention that is non-restrictive 'because choices are not blocked, fenced off, or significantly burdened' (Thaler and Sunstein, 2008: 5).

Nudges, in the words of Thaler and Sunstein, operate through the design of the 'choice architecture' structuring a situation or context within which a choice is made – in other words, the spatial, structural and interaction-based conditions in which choices are laid out and presented to the chooser. The designers of such arrangements, called 'choice architects', can apparently help improve aspects of our daily lives through the design of more user-friendly environments. Thaler and Sunstein cite not only the iPod and iPhone as examples of such a design, but also the arrangement of products on a supermarket shelf or in a cafeteria (with more healthy foods often placed at eye level and confectionery products placed lower down in order to nudge people towards a more healthy diet), or the layout of subscription forms and the design of registration processes (such as forms for organ donation or registration forms for some occupational retirement plans in the US, where a shift from opt-in to opt-out – from explicit consent to explicit refusal – would be a simple intervention in the choice architecture that could improve the 'well-being' of large numbers of people).

The gist of nudging is thus the idea of scrutinising and purposefully arranging choice situations in terms of their spatial and interactive architecture in order to *statistically influence decisions*. Nudge thinking comes from a behaviourist perspective that is geared towards managing the flows of users and individuals at crossroads in their daily lives. The rationality of nudge thinking focuses on the statistical average of large numbers of people in situations where many of us face similar choice situations. Nudge thinking tends not to focus on the singular opinions or mindset of each particular individual who might (or should) be convinced, educated or even manipulated towards a certain goal. Instead, nudging adopts a statistical perspective on large cohorts of choosers and their average behaviour as a product of a choice architecture, rather than of reasons and deliberation.

While this principle of nudging easily reminds us of strategies for the subtle manipulation of people – for instance, to the benefit of an economic actor – Thaler and Sunstein (and many others) insist that a behavioural intervention is only a genuine nudge when it influences choosers 'in their best interest' (Thaler and Sunstein, 2008: 9). Beyond the obvious questions of how, or on what grounds, to determine anyone's 'best interest' (and who does the determining), there is simply no inherent, conceptually evident connection between the idea of purposefully arranging choice architectures and doing this in choosers' best interests. In other words, why should the principles, methods and rationale of nudging not be used to encourage detrimental choices as well? Or choices that reflect corporate interests rather than a user's 'best' interest?

This ethical ambiguity (or moral flexibility) of nudging is particularly evident when it comes to the online world. Acknowledging the principle of nudging is key to understanding the designs of many graphical user interfaces, from individual dialogue boxes or menu and sidebar layouts to whole usage flows. For example, some users of Gmail, Google's free email service, will be familiar with the dialogue box that pops up from time to time after they log in, asking them to enter their phone number if they have not already registered it with their account. The dialogue box (see Figure 3.1) suggests that providing your telephone number serves as an additional measure to 'secure' your account against password loss and unauthorised access. However, it is much more likely that what really drives Google to use this aggressive form of nudging is its interest in phone numbers as highly effective personal identifiers. Through phone numbers, it is possible to interlink all the profiles and accounts that a user has set up on various networks – emailing platforms, messaging apps, dating apps, newspaper subscriptions, vehicle hire, booking and banking services and so on.

Figure 3.1: Screenshot showing Google account sign-in phone number nudge

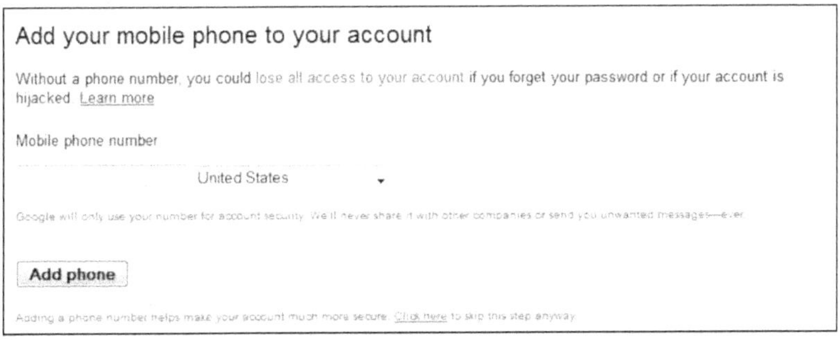

Source: Rainer Mühlhoff, screenshot from google.com, 2011

The deceptiveness of this dialogue box becomes clearer when we look at its graphical design more closely. Following the nudging credo, the option of *not* entering the telephone number *is* available (the design is not in theory being restrictive here) but *well hidden*. The link for this is a generically titled 'Click here' hyperlink placed far down in the grey, small-print (and hence, hard-to-read) text. Any usability expert or UX designer will tell you that this is actually the worst way of presenting a real choice. It is a truism to say that we tend not to read through the text contained in a dialogue box but intuitively search for 'clickable' objects. If the dialogue box were to include two adjacent buttons with the choice 'Enter phone number' or 'Not now', the probability that users would enter their phone number would dramatically decrease compared to the present more deceptive design, which is geared towards nudging users to make a particular 'choice'.

This is just one example from the extensive toolbox of design tricks employed by Big Tech to channel the flow of millions of users in corporately preferred directions. The business reality of usability and UX design practices today is primarily shaped by the aim of conceiving every dialogue box, interaction screen and interface as an instrument for influencing choices. I use the term 'interface nudges' to refer to situations in which nudge thinking is manifested in the design of a user interface as choice arrangement. Interaction flows, when viewed through the lens of interface nudges, are actually fine-grained sequences of nudges. Every screen that offers us the opportunity to click on this or that button, swipe left or right, continue or stop watching, is optimised in an application of behavioural science to exploit pre-reflective, subconscious, hard-to-resist mechanisms that frame choices in such a way that statistically averaged user behaviour can be steered in a particular direction. Nudge thinking as design methodology thus becomes the defining stylistic feature of how to interact with clients and users in the public as well as the online realms.

As mentioned before, in the field of public policy, an intervention qualifies as a proper nudge only if it increases the subjective 'well-being' of people in an empirically verifiable way. Interface nudges, on the other hand, are ambivalent with regard to the users' interests, to say the least, because they are clearly being used to advance the economic interests of the platform company, including their interest in capturing more data. I still refer to this as 'nudging' since there are overwhelming similarities between interface nudges and classic policy nudges. In both cases, the focus is on minimal interventions and the quantifiability of the effects; in both cases, choice architectures are subject to design considerations with the goal of prestructuring the decisions of many. In both cases, the designers have a clear idea of which option they want the average user to choose. Again, in both cases, no techniques of persuasion or argumentation are employed, but rather a shaping of the 'option space' that is intentionally biased towards certain predefined pathways. In both cases, the designers thus enjoy a greater position of power in relation to users. The only difference compared to nudging as a public policy tool is that proponents of the latter insist that this tool be used only for supposedly benign or 'paternalistic' ends.

The Web as a real-time behavioural laboratory

It is central to nudging that the effects on average user behaviour of minimal changes to the choice arrangement can be empirically verified. This raises the question of how this statistical average can be measured and how the efficacy of interface nudges can be quantified. Determining the 'quality' of an interface design has always been a major topic in the field of professional web development. A traditional approach has been to conduct so-called usability experiments, in which a group of randomly selected people use a website or software to perform certain tasks while the developers passively 'look over their shoulders'. Often, different variants of a design are compared in this way in order to 'empirically' decide which of them leads more often and more quickly to a desired result for a small sample of users. This technique is still used today, but its importance is waning because front-end developers now tend to use traffic analytics services in most cases.

Traffic analytics services are tracking services built into websites, smartphone apps or desktop applications. Usually, these services register the clicks and movements of the user through a website or app and transmit these data to a server so that they can be evaluated statistically. Most web analytics services can detect recurring user sessions on the same website (when the same user returns to a website later) and can match activities by the same user on different websites or apps. The market leader in traffic analytics services for websites has for a long while been Google Analytics, a free service from Google for web developers. Launched in 2005, Google

Analytics is currently used by 56 per cent of the 10 million most visited websites worldwide, according to statistics from the W3 Consortium (W3Techs.com, 2024). To enable Google Analytics on a specific website, web developers merely have to embed a short piece of code provided by Google in the HTML header of their site. This code then runs in the visitor's browser and uses JavaScript to record clicks and movements before linking with a Google server to upload these data. For website operators, Google Analytics provides a dashboard and various functions for visualising and analysing the collected usage data in real time. Subscribers can see how many visitors call up the individual sections and subpages of their site, how this is distributed over particular times of day and by visitors' geographical regions, and how this is affected by user attributes such as gender, age, income, interest profiles and the like.

In addition to these per-subsite statistics, tools like Google Analytics can also visualise user flows between subsections of a website, or the specific path of subsequent transitions between subpages and particular elements of content. This is of huge interest in a marketing context as it enables an analysis of so-called conversion funnels. These are specific sequences of steps in a website or app's usage flow that culminate in a predefined goal, such as completing a purchase. In the case of an online store, a conversion funnel might include the path from a landing page to a product search, and from there to viewing detailed product information, adding items to the shopping cart, navigating the checkout by following various intermediate steps like shipping dates and payment options, all the way to the eventual goal of a completed purchase. At each of these intermediate steps, a certain proportion of users typically 'drops out', transforming the chain of intermediate steps into a 'funnel' that gradually narrows.

A quantitative analysis of such conversion funnels is considered a key resource in identifying the steps with the highest dropout rate to optimise and improve an app or website for the average user. Besides content-related issues that make users drop out at various stages (such as quality of the offering, prices, shipping or payment methods), factors relating to a site's usability could also be responsible for heightened friction at particular stages. This is the point where HCI design steps in, with its quantitative approach to user behaviour. Optimising the conversion rate at every stage in a usage flow means conceiving of each of these stages as a choice architecture in which the choice not to leave the site but to proceed further in the direction of the site operator's goal is made to be statistically more likely to occur.

But how could the real-time laboratory of traffic analytics be used to come up with new design ideas to 'improve' those choice architectures? There is a data-driven solution for this that dovetails neatly with traffic analytics called 'A/B testing'. An A/B test is a quantitative method for comparing the performance of different design variations under real-world conditions. In web analytics, this is achieved through a mechanism on the web server that allows designers to provide different variants (A, B, C, D and so on) of the same website (or subpage), and the web server then randomly assigns each visitor to a group that will see one of these variants. As a result, these variants can be compared in terms of all the UX metrics (including conversion rates), each measured separately for the different groups. Variants in A/B testing typically involve only minor tweaks to the original (or baseline) design, such as different representations of price tags in an online shop, variations in the size or colour of buttons, or the spatial and visual arrangement of information (as mentioned earlier in the example of Gmail's phone number nudge). Because a website in real-world operation typically has a large number of users, and these users, unlike in a classic usability test, have no knowledge that they are participating in a test, A/B tests yield statistically highly significant quantitative results to aid in the assessment of design variations.

Through A/B testing, the interface nudging described in the previous section gains a robust empirical foundation. Each stage in an HCI flow can be considered as an action that unfolds in a choice arrangement, which in turn can be 'optimised' in the paradigm of nudging. Through the near-ubiquitous use of tracking techniques, the internet becomes a *real-time behavioural laboratory* to measure and model in detail how users react to various visual stimuli. Due to the convergence of design practices and big-data analytics, HCI and UX design are now fully data-driven disciplines.

Of course, there is no reason to assume that the data collected through traffic analytics services that are freely provided by companies such as Google are only used for design optimisation. Why is Google in fact interested in offering the Google Analytics service for free? The meticulous tracking of users as they navigate websites, providing a resolution as fine-grained as registering single clicks on website elements, measuring how far down one scrolls and the time one spends on various bits of content goes far beyond the data resolution provided by Google Search's click-tracking mechanism outlined in Chapter 2. Google Analytics is thus a clever way of extending Google's tracking mechanisms beyond the search result page to large parts of the internet. Google Analytics is actually an ideal hack for harnessing the involuntary complicity of web developers in order to establish a tracking infrastructure whose reach extends to millions of external websites. Monetising the tracking data elsewhere – for instance, through the behavioural profiling of users for advertising and risk-scoring purposes – is *the* business model of Google Analytics (whose parent company is, after all, the largest provider of personalised advertising on the Web). As we can see, there is a deep connection between interface design, data capture and the business strategies of platform capitalism. The interface designs we face every day are woven by capitalist interests and big-data methodology.

There is, by the way, no compelling reason for website-tracking services to be centralised. There are many available alternatives to Google Analytics that run on a website's own server without sending the data to a platform company like Google. The prevalence of centralised tracking services stems from a combination of convenience and a lack of care for the broader societal implications on the part of designers and developers, who effectively become accomplices to Google, the internet's largest data-capturing entity.

Sealed surfaces and the demobilisation of instrumentality

Interface nudges are intentional but at the same time subtle and implicit ways of steering and influencing user behaviour in digital environments. Tracking and web analytics operate covertly beneath the surfaces of graphical user interfaces to measure and predict how users react to specific design architectures and stimuli. These two parts of my argument in this chapter present the more covert aspects of what I term the 'design culture' of digital counter-enlightenment. Turning to a third aspect, we will now ascend from underground to the visible surface of contemporary design practices. Looking at various examples from the area of UX design, I will examine how contemporary design envisions the relationship between users and devices. Overall, I will argue that these design practices demobilise the instrumental use of reasoning about devices in favour of a perception of those devices as black boxes and active agents.

Very generally, UX design is a field of knowledge and practice on the borderlands where marketing, design and technology meet, whereby the design of user interfaces, menu navigation, the haptic and material design of devices, usage processes and application scenarios become the subject of systematised and mostly industrial strategies. In today's tech landscape, expertise in UX design is indispensable for modern technology companies because it plays a crucial role in introducing new digital products to the market by seamlessly integrating them into users' daily lives and forming user habits. UX design therefore has both market strategic value and cultural significance.

The term 'user experience', like the related term 'usability', refers to a subsidiary aspect of HCI. The ISO standard 9241–210 from 2010 defines UX as the totality of 'perceptions and responses resulting from the use and/or anticipated use of a product, system or service' (ISO, 2010). More specifically, the norm states:

> User experience includes all the users' emotions, beliefs, preferences, perceptions, physical and psychological responses, behaviours and accomplishments that occur before, during and after use. [...] User

experience is a consequence of brand image, presentation, functionality, system performance, interactive behaviour and assistive capabilities of the interactive system, the user's internal and physical state resulting from prior experiences, attitudes, skills and personality, and the context of use. (ISO, 2010)

The term 'user experience' should be distinguished from an earlier and more narrowly defined paradigm of usability. There is also a standard (ISO 9241–11 from 1998) that defines usability as the degree to which a technical service can be used to achieve a certain goal 'effectively', 'efficiently' and 'satisfactorily' (ISO, 1998). In the usability perspective, *instrumental* logic prevails: usability is measured by how easily someone can use an interactive service as a tool to complete certain tasks. UX, in comparison, seems a slightly diffuse and fuzzy concept because it takes into account subjective aspects such as 'fun', 'sensuous qualities' and 'enjoyment' in connection with an interactive product. Instead of the instrumental focus that highlights how useful the tool is in terms of representing an effective means to an end, the UX perspective takes a holistic approach that includes the entire world of feeling and perception to measure the quality of product and design decisions.

The term 'user experience' harks back to the Californian design pioneer Donald Norman, who made a name for himself with his 1988 monograph *The Psychology of Everyday Things* (Norman, 1988). In the early 1990s, Norman worked at the Apple company as a 'user experience architect' – the first time that term had been used in a job title. It was not until the 2000s that UX emerged as a dominant paradigm, which can be seen in some sense as a counter-movement to the usability paradigm, reframing human–computer relationships as not purely instrumental. Over the last decade, finally, UX has seen massive commercial adaptation, as a result of which the distinction between UX and usability has become blurred.

As part of this focus on UX design, a trend of 'sealed surfaces' is emerging in contemporary digital media culture – the observation that the design of digital devices often conceals the technical nature of the device beneath closed surfaces, which manifest as user interfaces. The design of human–machine interfaces is increasingly involving the use of design features that hinder users from gaining insights into the technical workings of devices and applications. The technical reality that is the essence of these objects is represented to users both haptically and terminologically in the guise of simulated (often anthropomorphised and naturalistic) forms of interaction.

For example, on our computers, we place 'documents' on the 'desktop' or 'throw' them in the 'rubbish bin'. We 'search' our computers, put 'files' in 'folders' or 'share' them with someone else through the 'cloud'. Design decisions like these wordings and depictions have a methodological background deeply entrenched in Apple's corporate philosophy and market strategy, which have been defining the industry standard of usability and UX ever since the 1980s. In what has often been termed 'skeuomorphic' design, Apple has long been trying to graphically simulate on the computer or smartphone screen the objects users would find in the real world of their offices – all the way to simulating wooden or leather textures on surfaces, or providing faux 3D effects when a button is pressed (Irish, 2022). Skeuomorphism is a design principle that involves incorporating design elements or features into a product that mimic the characteristics of real-world objects, even if they are not functionally necessary in the digital context. That is, skeuomorphism *simulates* on a digital screen physical behaviour, material textures or mechanical processes that reference the real-world counterpart of a digital object, but which are not needed for the device to function. Designing these features involves dedicated and sophisticated coding and extra computational resources on the device for something that is, from a functional perspective, a mere decorative add-on. From a sociotechnical perspective, however, these 'decorations' of simulated physical reality play a role that deeply shapes users' relationships to the technological artefacts, as they constitute how we perceive and understand the digital objects that are per se abstract and non-intuitive to most users.[4]

Even now, when skeuomorphism as a design trend has passed its peak, giving way to what since the mid-2010s has been touted as 'flat' design (Turner, 2014), the design of touch-screen interaction follows a similar path. When touching and moving elements around on a screen, interfaces are designed so that these graphic elements simulate the behaviour of their real-world, physical counterparts: scrollable areas simulate mass and inertia once set in motion, causing scrolling to continue while gradually slowing down; movable objects like 'icons' or 'windows' are simulated to 'snap' into grid points as if attracted by a grid of magnets; the act of refreshing messages or email inbox lists entails users following a 'pull to refresh' interaction design, whereby you pull the list beyond its limit against the elastic force of what intuitively feels like a spring.[5]

In addition to what we see on the screen, the new generation of smartphones no longer has a battery cover or screws that could provide access to their inner workings. Smartphones physically present a sealed, smooth – often overly smooth – surface. In all these aspects, the trend of sealed surfaces comes with immense efforts to make the devices simulate naturalistic behaviour at their surfaces that they do not actually possess. The actual logics of the processing of bits and bytes in the digital circuits within the device are made intuitively accessible to the end user by means of a sophisticated graphical extra layer that is supposed to somehow 'fit' people's habits and perceptions, drawing a strong line of demarcation between the inner and the outer interactive workings of the device.

Figure 3.2: Vacuum cleaner nozzle switch icons. Left: old design; right: new design

Source: Rainer Mühlhoff

Another facet of the sealing surfaces of digital devices concerns the strategically demobilised *instrumentality of digital devices*. Let us take a brief excursion outside the field of the digital by using the intuitively simpler example of a vacuum cleaner nozzle. This type of nozzle often has a switch that can be activated by foot and, for instance, displays the symbol of a 'bristly' nozzle in one position, while another position shows the symbol of a 'smooth' one (see Figure 3.2, left side). These positions are used depending on whether one wants to vacuum a smooth surface such as a tiled floor or a soft surface such as a carpet. At some time in the last two decades, there was a turning point. A different way of illustrating the switch became more common. In this new design, the nozzle is not shown any more, but the user can choose between a tiled and a carpeted floor (see Figure 3.2, right side).

   This change from the first to the second design is crucial. In the first case, the switch communicates a property of the nozzle, namely whether it is bristly or smooth. Communication with the user revolves around the mechanical configuration of the vacuum cleaner; hence, the device can be used and configured by the user as a *tool* for an indefinite purpose that is fully at the user's discretion. In the second case, the switch is designed so that the user has to *communicate* the type of flooring to the device. Communication

Figure 3.3: Two different microwave interface designs

Curry / Cheese
Pasta / Casserole
Fish / Vegetables
Jacket potatoes
Potato products
Chicken
Pizza

Source: Rainer Mühlhoff

with the user revolves around the mechanical properties of the floor and the user does not get involved in determining the mechanical property of the device that would best serve this purpose. Similar observations could be made about user interfaces when it comes to microwaves, washing machines and various other household appliances. With a microwave (see Figure 3.3), there is a design variant with only two knobs for setting the radiation power (in watts) and the duration (in minutes or seconds), and another variant with numerous buttons like 'Fish', 'Chicken', 'Pasta', 'Popcorn' and so on. The second design does not reveal that all these fancy buttons effectively do is make a selection over the same two-parameter space of power and duration, based on a hidden recipe. The second design thus *seals away* technical information and control options from the user, allegedly making things more convenient by providing an interface that seems again to fit more immediately into the user's everyday reality.

Observations like this point to a cultural demobilisation of instrumentality (that is, tool-likeness) in the way users are encouraged to perceive and relate to technical artefacts. This demobilisation of instrumentality is part of what I call the 'trend of sealed surfaces'. It concerns the way in which a device's relationship to its intended use is interactively, discursively and symbolically represented. Since the end of the 20th century there has been a trend popularising forms of interaction in which users are not expected to comprehend the mechanical or structural make-up of a device in order to be able to deploy it in a means-to-an-end relationship as they see fit. This step, involving processes of thinking and understanding, and as a result of which the device could in principle be used for a non-predefined purpose based on comprehension of its mechanical configuration, is denied to the user in the more recent design culture. Interaction with the device then revolves around directly negotiating the end purpose – what needs to be hoovered or heated, for instance – but without according autonomy and authority to the user in terms of how to perform this operation. It is left to the device and remains opaque how this purpose is fulfilled.

The more recent vacuum cleaner nozzle design seems to presuppose that a large proportion of users do not want to think about which mechanical feature of a nozzle is best suited for which type of surface. The paternalistic design emerging from this rationale is touted by manufacturers as more convenient, more intuitive and more accessible. Similar observations can be made about many digital interface designs as well. For example, the Windows 3.11 operating system, popular in the early 1990s, had a File Manager (see Figure 3.4, top). The program graphically represented the various storage devices (hard drives, floppy-disk drives) built into the physical computer and, in a tree-like depiction, the directories and files stored on these devices' file systems. In Windows 95 and subsequent versions, the File Manager was replaced by Windows Explorer, which essentially had the same functionality but featured additional elements that allowed for a process-driven, 'exploratory' access to files (see Figure 3.4, bottom). For instance, particularly 'relevant' folders like Pictures or Documents were displayed in a central location, making it possible for a user to quickly select them without needing to know their structural location on the device's file system.

In a parallel development, in the 1980s the Apple Finder, with its characteristic user interface, was devised by Apple engineers. The Finder represents another paradigm shift in interaction design: it focuses on accessing objects through search interactions. Search interaction brings a dialogical principle of question and answer to the fore, whereby the representation of objects is even further uncoupled from the structural location of those objects on the computer's file system. A search-based interaction can even yield heterogeneous lists of items not limited to files, but also including applications, bookmarks or web pages, which serves to muddy the user's perception of the ontology of distinct types and categories of objects on (or in) the computer.

Core software interfaces of an operating system, such as those for accessing files and directories, are crucial elements that shape how users acquire practical knowledge and general perceptions about computers and what is 'inside' them. The terminological evolution from 'managing' to 'exploring' to 'finding' is highly noticeable when we compare file manager interfaces. This progression is not just about a change in names; rather, it represents different user subjectivities that are supported and inspired by the respective interaction designs. In this way, we see how users' consciousness of digital devices can be fundamentally shaped in different ways that either emphasise or avoid an instrumental understanding of what the device does.

Figure 3.4: Top: screenshot of File Manager (Microsoft Windows 3.1); bottom: screenshot of Windows Explorer (Microsoft Windows 8)

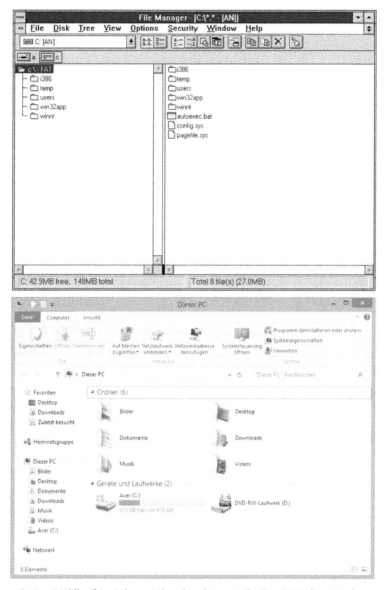

Source: Rainer Mühlhoff (top); https://de.wikipedia.org/wiki/Datei:Windows-Explorer_ (Windows_8.1).png (bottom)

The trend of sealed surfaces thus includes two aspects: first, users are not supposed to – or are assumed not to want to – use a (digital) device instrumentally, meaning as a tool with a generally indefinite list of purposes, which the user can operate through conscious reflection and knowledge of the device's technical characteristics. The prevalent picture of the 'average' user seems to be of a person who does not act out of rational comprehension but intuitively and with little foresight, step by step reacting to stimuli and affordances as they apprehend them. Second, users are assumed to have no desire to comprehend the internal workings of a device as a technological and artificial entity. The design of digital artefacts strives to conceal the 'technical' and 'artificial' beneath a surface that uses actively simulated outward behaviours to suggest a 'natural' and 'intuitive' interaction with the device that seamlessly merges with our everyday experience of material objects.[6]

As a consequence, this designed demobilisation of an instrumental relationship to digital devices paints those devices as *active* agents in relation to their human users. As much as their innards are sealed and hidden from view, and as much as their externally perceptible operations simulate physical and human behaviours in the real world, these devices are ideal vehicles for the ascription of qualities like volition and agency. A UX design company like Apple is fully aware of this when it directs its staff in Apple stores not to use language such as 'the application crashed' or 'the device hangs', but instead to say, 'the application unexpectedly quit' or 'the device does not respond'.[7] It appears to be strategically advantageous for commercial and corporate interests to cultivate an aura around digital devices akin to that of a living being. As little as one would generally 'use' a living being instrumentally do we usually dare to explore its innards. If an animate being (like a pet) does not do what we expect or want it to do, we need to deal with its volition and can even end up feeling helpless or overwhelmed. With ever more digital devices now being fuelled by AI technology, this framing of devices as active agents easily leads to a kind of AI animism: the attribution or projection of human qualities such as consciousness, emotions, intentions or a soul onto AI devices (see Proctor, 2018).

Subjectivity of digital counter-enlightenment

Within the commercially driven culture of contemporary digital interface design lies a set of ideas about what users are presumed to expect, imagine and be capable of in relation to digital devices. This ideology forms an implicit anthropology that relies on at least three pervasive assumptions:

1. User behaviour is not based on instrumental reasoning and technical comprehension, but is rather driven by impulses, social dynamics and pre-reflective stimuli. The mechanisms governing user behaviour can be statistically measured and predicted to optimise the design of HCI.

2. Users are not interested in gaining insights into the technical structures, algorithms and mechanics of their digital devices; when confronted

with technical details, they quickly feel helpless, and hence such details should be withheld from the average user.
3. Users tend to buy into the framing of digital devices as active agents that we do not command, but with which we interactively relate.

I refer to the mindset that relies on these assumptions as the ideology of digital counter-enlightenment. I use the term 'ideology' because commercial discourses and practices tend to naturalise this mindset as an objective fact. Challenging this naturalising tendency, the approach of this book is to analyse the technological culture that emerges from the powerful design practices of digital counter-enlightenment as a constellation of *technological subjectification* (see Chapter 4). This means, heuristically, that the design of human–machine interaction has a formative effect on the users of digital devices and influences their shared discursive and cultural practices of perceiving and relating to these products. Users who project agency onto their devices, as well as users who feel less competent or interested in acquiring common-sense technical knowledge about their devices, represent a product and effect of this design culture.

It is the basic selling point of usability and UX that it is not people who should adapt to technical devices, but technology that should adapt to users' abilities and needs. Of course, this idea rings true and has influenced many interface designs for the better. But this assumption is wrong in terms of social theory because it suggests a one-sided process of adapting technology to supposedly fixed, universal and natural human dispositions. Interaction with technology shapes habits, modes of perception and physical intuitions in dealing with devices; it shapes the way in which users experience, discuss and relate to technical artefacts. The degree to which we understand our technical devices is a result of technological and cultural influences. In this sense, interface design is a key factor governing how subjectivities in the context of digital culture are constituted – that is, how we perceive, comprehend, make use of and pragmatically relate to digital artefacts.

I have described the ideological regime that shapes the form of subjectivity induced by contemporary mainstream technology design as 'counter-enlightenment'. This is to highlight the structural and cultural rather than individualistic and naturalistic framing we need to adopt in a critical approach. I do not refer to a failure or predisposition on the part of some users to avoid reflecting on, understanding or otherwise engaging in instrumental reasoning. Rather, my aim here is to analyse the structural conditions that co-create our shared consciousness in relation to digital devices. In this shared consciousness, instrumentality is demobilised and the fiction of the device as an active agent is advanced as a product of design practices to which most of us are exposed on a daily basis. Together with the strategic exploitation of stimulus–response behavioural patterns, all three aspects of what I call an anthropological imaginary of the user in digital counter-enlightenment serve corporate interests. These aspects work together to make us available to corporate interests as data producers attached (and sometimes addicted) to our digital devices. The next chapter looks in more detail at the notion of subjectivity and describes the production of user subjectivity as a power strategy that lies at the heart of the operations of AI as it is manifest in *Human-Aided* AI systems.

4

Subjectivity and Power
in the Ethics of AI

AI and data companies such as Amazon, Google, Meta, Alibaba, Tencent and Baidu represent an accumulation and centralisation of economic and political power on a scale unprecedented in today's (post)industrial capitalist world. Given the amount of (personal) information about nearly all human and societal affairs that these companies accumulate, their enormous market capitalisation, their quasi-monopoly status in wide market areas and their immense political influence worldwide, we might think of them as new sovereign power poles, potentially on a par with the sovereign power of states.

As the previous chapters have outlined, AI today is based on the worldwide, massive harnessing of users and workers as data producers and cognitive subunits of computational systems. The Human-Aided AI paradigm makes visible how AI technology relies on the continued collaboration of billions of people, thus establishing a sociotechnical perspective on AI. Turning to the issue of the power of AI apparatuses, the notion of Human-Aided AI therefore provides for a more extended analysis, envisioning the power of AI (companies) not only as a phenomenon of centralisation, but also as a social and cultural arrangement in which power disperses and emerges from myriad smaller relations between users and their devices. A critique of the power of AI should thus raise questions regarding the ways in which we are not only potential victims or beneficiaries of the power accumulation of AI companies that are external to us, but rather *active* agents in the power of AI apparatuses. If we look at the large-scale capture of users as data producers, including users in more privileged and affluent positions within liberal societies, we must concede that a majority of us participate in and contribute to AI apparatuses. This participation occurs to varying degrees voluntarily, or at least is self-perceived to be a free choice, often seen as advantageous, rational, pleasurable or simply enjoyable.

Along these lines, Human-Aided AI engenders a perspective on the ethics of AI that factors in how the habits, perceptions, behaviours and

75

subjectivities of users of digital services provide the foundation for AI in its current form to exist. Human-Aided AI raises the questions: How does AI depend on our very personal, everyday, affective and reflexive relationships with networked media? What makes us so available to use the commercial services and interfaces of AI companies on such a large (in truth, global) scale? What makes us so attached to these devices that almost all of us uniformly and seemingly of our own free will – though not necessarily consciously – contribute to AI apparatuses and their unprecedented profitability as business models?

The critical philosophical paradigm of Human-Aided AI opens up space for identifying the phenomenon of our voluntary participation as the locus for critique and raises questions about our own responsibility in relation to AI systems. Given that the current proliferation of AI and big-data technologies is not only beneficial to individuals and society but also comes with certain risks or even harms, the question of our own *in*voluntary collaboration in contemporary AI systems takes on an ethical significance. In fact, given how many individuals and groups can easily find themselves victims of social sorting, algorithmic discrimination, manipulation and the reinforcing of social inequalities (Barocas and Selbst, 2016; see also O'Neil, 2016; Eubanks, 2017; Mühlhoff, 2020b, 2023a), our daily data-sharing practices and implicit collaboration with Human-Aided AI systems take on a paradoxical flavour. From both an individual and a collective viewpoint, the wider consequences of our collaborative behaviour arguably go against our own interests, including equal treatment, non-discrimination and maintenance of our autonomy.

One of the prime challenges in the ethics of AI is hence to explain and ethically investigate a paradox between individual behaviour and collective effects: as users of technology who widely share certain behaviours, perceptions and rationalities with regard to digital services, we are collectively enacting and enabling systems (Human-Aided AI systems, AI capitalism, AI extractivism) that on a global scale have harmful and detrimental effects on potentially any one of us, as well as on social groups and wider societies. In some way, this makes the ethical and political issues thrown up by AI similar in kind to the primary ethical and political challenge of our times: the climate crisis. Both problems need to be addressed at a systemic level. That is, an ethics of AI (and, equally, ethical approaches to the problem of climate change) must be pursued as a questioning of a *structural* and *systemic* constellation (see van de Poel et al, 2012). As a consequence, any ethical approaches that focus on our actions, moral deliberations and decision-making *as individuals* are toothless in relation to the globally interdependent and collectively enabled constellations in question.

In this book, I seek an ethics of collective responsibility that informs and unleashes political action as a path out of the impasse of classical ethical and philosophical approaches to the sociotechnical constellation that is AI. But how we reach this destination is by no means straightforward and relies on two intermediate theoretical steps: an analysis of power and of subjectivity in the context of Human-Aided AI. In the present chapter, after (re)conceiving AI systems as networked power apparatuses, a theory of subjectivity and subjectification in the tradition of poststructuralist critique will be used to provide the link between sociotechnical apparatuses as a whole and individual actions, perceptions, behaviours, evaluations and so on. The concept of subjectification allows us to theorise the relationship between individuals and the digital culture of AI capitalism as a mutually *constitutive* one, whereby constant exposure to the interfaces, services, use cases, corporate rhetorics and rationalities of the networked media landscape shapes our subjectivity as users and data producers and thereby reproduces the human cognitive agents that are the vital resource for AI business models.

The reproduction of Human-Aided AI

The structural and systemic constellation we need to scrutinise and politicise in the ethics of AI is *not* external to us as users, workers and citizens. In particular, the users of networked digital devices in the Global North who are confronted only with the 'soft power' of data capturing are to a certain degree accomplices in the power of AI apparatuses.[1] An ethics of AI that takes a systemic perspective therefore needs to 'ethicise' and 'politicise' the habits, behaviours, perceptions and rationalities of the users of AI systems, not as a result of individual decision-making but of systemic co-constitution.

Such an analysis of the systemic co-constitution of individual habits, behaviours, perceptions and rationalities in the service of large-scale power apparatuses is enabled by the concepts of 'subjectification' and 'subjectivity' as conceptualised by Michel Foucault, Judith Butler, Gilles Deleuze, Felix Guattari and others. In my specific use of the terms, 'subjectification' refers to the social and relational processes of constituting individuals as agents of a subjectivity. 'Subjectivity', in turn, is a form of reflexive relationship to oneself, to others and to the world.[2] Crucially, this self-reflexivity is constituted as a product of a social arrangement or power apparatus that includes its own norms, categories, practices and rationalities. Subjectivity, as Foucault states, is 'the way in which the subject experiences [it]self in a game of truth, in which [it] relates to [it]self' (Foucault, 1998a: 461). Subjectivity thus emerges in a field of relations to others at the same time as it is mediated by (and created through exposure to) social discourses, practices, narratives, bodies of knowledge, infrastructures, media operations and a host of other social practices and phenomena we engage in every day.

Applied to the present case, the idea underlying subjectification is that contemporary digital culture, which is significantly influenced by commercial actors through interface designs, use cases and corporate messaging, plays a crucial role in shaping user subjectivity, and that this shaping of user subjectivity is a central mechanism in *reproducing the human resources necessary to maintain contemporary AI systems that rely on human collaboration*. Insofar as AI systems are built from human resources in addition to semiconductors, wires and rare materials, inducing a 'suitable' subjectivity of users is a main function of the power of AI apparatuses to the extent that it serves the (re)production of complicit users.[3]

The full strength of the concept of subjectivity in the context of a critical philosophy comes to the fore chiefly in relation to the notion of power in Foucault's writings.[4] For Foucault, power is a relational and decentralised phenomenon, dispersed throughout a social (or sociotechnical) arrangement, involving everyone and everything.[5] Enabling a structural account of power, the main questions in this approach are not about who has power and who does not have it. Given that everyone has power to varying degrees in relation to other people, the most pressing question is: What kind of decentralised interplay of power emerges from the myriad micro power relations within a social arrangement, and what kind of more rigid structures become stabilised in that interplay? (See also Foucault, 1978: 92–102.) In particular, this approach starts from the premise that every individual – considered as a subject – is by default an *active* and *powerful* part of a social arrangement. Each individual wields a small portion of power immanent in the power of the overarching system; or, to put this the other way round, the power of the overarching system – for instance, of a Human-Aided AI system – emerges from the interplay of myriad micro power relations involving, among others, the power to act of users who are subjects in that very apparatus.

Why is it important to add 'who are subjects in that very apparatus'? This is the clue to the immanence-based philosophy of power and subjectivity established here in reference to Foucault and Deleuze (see Deleuze, 1988a). The overarching apparatus is *enacted through the actions of its subjects*; its total power is constituted by the power of those subjects and, conversely, that overarching apparatus produces the subjectivity of its subjects that shapes and motivates their actions, perceptions and evaluations. The relation of part (subject) to the whole (apparatus) is therefore one of immanence. The power apparatus exists in nothing but the actions of the subjects, but in a non-cumulative way, as the whole has a backward-operating effect on the parts (in the form of producing subjectivity). The actions of the subjects are more or less shaped by the apparatus in such a way that the apparatus reproduces and perpetuates itself.[6] The term 'apparatus', or *dispositif* in Foucault's terminology, might denote any social, economic, political or societal configuration that involves discourses, practices, norms, bodies of knowledge, technologies, media devices, institutions, economic structures, schools, police, jurisdiction and penal systems, for instance.[7] In order to reproduce itself, this social apparatus 'produces' individuals as subjects – that is, as agents whose (self-perceived) 'free' actions, behaviours, desires, wishes and so on are roughly in line with the interests of the apparatus. As stated before, this constituting of subjectivity in each individual in the context of an overarching apparatus is referred to as 'subjectification' in Foucault. Subjectification is thus a function of a form of 'power which makes individuals subjects' (Foucault, 1982: 781), where the term 'subject' refers to the individual as the agent of a specific subjectivity.

Why is this approach a fertile one in a critical philosophy and ethics of AI? Because it provides a theory of the generally active and by default collaborative and co-creative role of users in AI systems that emphasises the non-volitional and non-deliberative nature of users' actions and behaviours. Merely by virtue of being a subject in our contemporary media culture, one is simultaneously contributing to maintaining the power apparatus, such as Human-Aided AI systems, through one's actions and perceptions – that is, through one's applying of the social discourse, communication practices and narratives in relation to our digital devices, knowledge practices, perceptions and rationalities to ourselves, as well as in relation to other subjects.[8] This philosophy of immanent power and its effects on subjectivity shows that not only do power relations act on the subjects as their targets and endpoints, at the same time they act *through* the subjects who contribute to perpetuating and stabilising the power apparatus to the extent that they exercise a complicit subjectivity.

This description of individuals – or users – as subjects of digital media and AI apparatuses is a long way from a deterministic and mechanistic, or even totalising, depiction of the social. On the contrary, the mutual dependency of subjects (who want to 'make sense' of themselves and the world and thereby rely on social discourses and practices of 'intelligibility' [see also Butler, 2004; Oksala, 2005]) and the overarching social arrangement (which tends to reproduce itself through the voluntary actions of its subjects) offers leeway for critique, interference, friction and resistance.[9] Subjects are roughly (though neither fully nor uniformly) shaped in their subjectivity by the overarching apparatus. There is still difference and disparity in personal motivations, perceptions and evaluations among the population of any social arrangement, showing that a form of power operating through subjectification is not a repressive, deterministic and uniformity-inducing one, but rather a purportedly 'humanising' power that nudges its subjects, instils economic incentives and gives many of its individuals (at least in the centre of society) the feeling of being treated as an individual and as a subject who exercises free will.[10] Understanding how, despite all degrees of (self-perceived) freedom and individuality, our motivations, perceptions, evaluations, desires and pleasures are *roughly* shaped by processes of subjectification is thus the prime avenue of critique and transformation of social arrangements that operate through the power of subjectification. Critique and resistance in such arrangements rely on self-critique to cultivate an agency that interrogates the constitution of our own subjectivity within the apparatus inasmuch as this might be co-fashioned by the interests of others.[11]

Applied to the context of Human-Aided AI, a critique of subjectivity enables us to understand how subjectivity is simultaneously a product of the digital media culture established through contemporary design practices, as well as an active agent in enacting and reproducing that very media culture as an apparatus of power. As we use our MacBooks, Android phones, Dropboxes, Gmail accounts, Instagram profiles, Alexas or Siris, our dating apps, fitness checkers and health trackers, we are 'accomplices' of AI. Many of us, especially in the middle classes of regions in the Global North, do not face AI as a potentially threatening, overpowering actor, but are involved in it through a close and seemingly beneficial symbiosis with digital devices and services that shape our subjectivity. Uncovering this codependence between AI and its users as a mechanism of power marks a self-reflexive and self-critical turn in the ethics of AI that is especially needed among the generally privileged groups that participate in academic discourse in the field.[12]

AI extractivism revisited

One message from Chapter 2 is that contemporary AI relies on the exploitation of at least two forms of power relations that need to be distinguished. The first form of power, that of *subordination*, makes use of global economic disparities to exploit precarious labour through click-work and sweatshop conditions in the service of AI. I call this power 'subordination' because it is built on stark power differentials in the global economic order (a product of colonial exploitation), reminiscent of sovereign 'power over' and a power that represses and dispossesses.[13] As my remarks on click-work in Chapter 2 bear out, this kind of power is not overly reliant on subtle or 'soft' strategies of subjectification. Click-workers are treated as software components and 'blank' human resources, and often seem driven to this type of labour through economic despair produced by inequalities that range all the way from Silicon Valley and other tech hubs through 'business process outsourcing' intermediaries to the workers' own communities and their employers' local hierarchies. Miceli et al (2020: 1) present an extensive analysis of this global chain of power relations, showing how 'the work of annotators is profoundly informed by the interests, values, and priorities of other actors above their station'. They argue that direct 'imposition', rather than the workers' 'subjectivity', should be a key analytical term in understanding the actions and perceptions of these click-workers.

The other form of power fuelling Human-Aided AI *captures* humans worldwide as cognitive subunits of AI systems. This form of power tends to operate at a local, micro level because it manifests in the relationship between users and digital interfaces or devices. Compared to subordination, capture as a form of power has been less debated in the ethics and critique of AI, but it is arguably even more pervasive since it affects nearly all internet users worldwide. In analysing the capture-based form of power, it is fertile to distinguish two related aspects of its form of operation: a situated, ephemeral aspect and a trans-situational, more stable one. First, networked interfaces comprise situated, momentary power differentials when users are tricked or subtly nudged by the design (and designers) of interfaces into producing data. Here, the psychological, often 'subcortical' tricks of nudging, dark patterns, UX design and addictive design (see Chapter 3) come into play – the ephemeral aspect of this form of power shapes the moment-by-moment, context-dependent behaviour of users who are embedded in relation to an interface as a result of visual and other cues that they receive. So, it is fruitful to think of networked interfaces as 'membranes' across which a disparity in power can be transformed into computing power.

The second feature of the power strategy of capture is its reproductive aspect. This is revealed when exposure to digital media and its diverse services creates subjectivities – that is, enduring attitudes, habits and perceptions, feelings of entitlement or practices of social interaction that can lead to lock-in effects in the form of a subject's or social group's attachment to specific digital services and the kinds of interaction they encourage. This temporal emphasis on an impact lasting over time characterises the phenomenon of user subjectification described earlier: being exposed to nudges, manipulative tricks, addictive design and ecosystem lock-in effects over time *creates* users who take decisions impulsively or intuitively as nudged by the digital 'choice architecture'; who resign themselves (whether consciously or not) to potential manipulation, embracing a symbiosis with commercially driven digital services and lock-in effects with respect to the ecosystems of specific companies' digital goods. This is nothing less than a politics of interface design that facilitates the inscription of digital services into the perceptions, habits, social skills, values, desires, pleasures and rationalities of billions of users across the world.

In this comparison between the subordinating and the capturing type of power involved in Human-Aided AI, it is clear that the capturing type does not operate in a hierarchical way, but instead operates more horizontally by wielding the instruments of manipulation and subjectification. Rather than forcing a user into a role as a data labourer, it shapes and mobilises the intrinsic motivations, rationalities, habits, communication patterns, desires and pleasures of individuals as 'free' subjects. While the subordinating power 'imposes interests, values, and priorities' of corporate actors (Miceli et al, 2020), the efficacy of the 'soft exploitation' of user capture is predicated on addressing subjects as people whose lives can be augmented, whose decisions can be eased and their social relations spiced up, and whose daily routines can be made more efficient. As a consequence, the form of power relying on user subjectification and capture is more difficult to grasp in its ethical implications because it makes everyday users simultaneously victims and co-perpetrators, lab rats and beneficiaries, products of subjectification and involuntary accomplices of AI systems that, from an aggregate perspective, exacerbate social inequalities and sustain systemic injustice.

AI extractivism relies on both these modes of power, along a spectrum from subordinating to subjectivating, from hierarchical to seemingly flat, from impositional to nudge-based forms of power. These modes of power, interwoven and mutually reinforcing, come together in the project of colonising *all* social relations in *all* areas of life worldwide through datafication.

Systemic critique and ethics beyond individualism

In highlighting the myriad tools that tech companies employ to capture ordinary users, including the comparatively privileged users from the Global North, as data producers and cognitive subunits, my emphasis is not so much on the fact that those users are victims of the design tricks of Big Tech (which, of course, they *are*). Rather, my emphasis is on our constitutive involvement in AI; on the fact that by way of subjectification, we are accomplices – that is, contributors to and beneficiaries of those apparatuses that capture, exploit and discriminate on a global scale. Because AI depends so much on daily users as involuntary data producers, the end result is that this imposes on its day-to-day users a collective responsibility with respect to the mechanisms of operation and the effects of this technology, including the exploitation of click-workers and the reaffirmation of global economic disparities and hierarchies that are partly a legacy of our colonial past.

Some commentators have characterised approaches that highlight the co-responsibility of users as a neoliberal strategy of 'responsibilisation' (Hache, 2007; Gray, 2009; Pyysiäinen et al, 2017). Responsibilisation is a tendency, particularly evident in behavioural science–based public policy approaches, to shift the political responsibilities of the state (such as preventing or minimising societal harm arising from AI companies' activities by passing appropriate regulation) onto the moral conduct of individuals. This is particularly prevalent in debates on climate change and sustainability in societies of the Global North: instead of governments and policy makers putting in place sufficient regulatory frameworks to tackle the issue at a systemic level, consumers are called upon to define and make their own sustainable consumption choices (see also Henkel et al, 2018). Similarly, in the case of digital technology, some researchers have argued that, for instance, the framing of 'privacy' as 'individual control' of data (see also Kröger et al, 2021) amounts to a strategy of responsibilisation of users with regard to the potential threats resulting from corporate data aggregation (see also Lisker, 2023). This is a strategy that distracts attention from more far-reaching political action and regulatory projects.

In transforming users from being potential victims of data capture and exploitation into collaborators with and accomplices to Human-Aided AI systems, my message is not that *individual* users should be held culpable and responsible. Human-Aided AI looks at AI systems as sociotechnical constellations. Consistent with critics of responsibilisation, I maintain that the responsibility for systemic arrangements can and should not be individualised and that incentives for behavioural change should not be the main component of a viable policy for systemic transformation (Gray, 2009; Pyysiäinen et al, 2017). While it is true that most users, including those in comparatively privileged positions, do share part of the responsibility for the global risks and harms of AI capitalism, this responsibility should be seen as a collectively owned responsibility (see Chapter 11 on the theoretical underpinnings of this argument). Users are co-responsible collectively, in a way that goes beyond the (false) notion that each user shares a *cumulative* part of that responsibility. To make sense of this, we must stop centring ethics around individual behaviour and instead look at the large-scale *seriality* of the behaviours, perceptions, attitudes and habits of many. To this end, critical analysis of subjectification is a main tool of a power-aware approach to ethics, with subjectification pointing to *shared* forms of subjectivity. Akin to how the 'imperial mode of living' of citizens in the Global North as a widely shared, potentially unconscious form of subjectivity is contributing to the current climate crisis (see also Brand and Wissen, 2021), the feasibility of AI and AI capitalism with its global and local effects relies on a sense of entitlement and attitude of neglect culturally encoded in the way most of us relate to digital media. The critical question is thus how the legal, political and technical feasibility of discriminatory and exploitative AI systems hinges, in part, on our more or less shared subjectivity.

The related ethical issue is therefore not that of the morally right behaviour of a particular moral agent (for example, a user). Individual agency is irrelevant in a constellation in which we interact with huge companies that are gathering and aggregating data on a mass scale. It is ineffective and misses the point if individuals decide not to use certain services or are tempted to hold themselves personally accountable for the consequences of AI that derive from the fact that *many* people in our societies are simultaneously using AI services without reflecting much on the implications of this trend. An ethics of AI should, rather, be formulated as an ethics of collective interests, systemic critique and collective responsibility-taking. The ethical and political issues arising from the systemic critique pursued in this book address the possible ways in which we could limit and control the power imbalances, societal ruptures and inequalities, structures of exploitation and infringement of autonomy, equality and privacy that are among the global and local effects of AI capitalism. An ethics of AI should thus mobilise collective responsibility that can be channelled into collective political action of users as political citizens and members of a global community. This emphasis on the collective is based on the analysis of power and subjectivity in Human-Aided AI as a *structural power*, where 'structure' refers to constellations of a dynamic stabilisation of roughly similar and parallel perceptions and behaviours between many subjects (see the Introduction). It is precisely our widely shared behaviours in sociotechnical apparatuses, orchestrated by subjectification through digital media interfaces, that enables AI to become a product of our shared subjectivity in contemporary media culture.

PART II

The Power of Prediction

AI systems as Prediction Machines

The previous part of this book developed a sociotechnical perspective on AI systems. This approach aims to foreground the profound conceptual and practical challenges to the ethics and critical philosophy of AI that arise from the fact that AI is not a technical object outside of us, but a media-cultural, economic and sociotechnical constellation that, in addition to corporations, financial markets and tech billionaires, involves billions of ordinary people as users and data workers. These considerations emphasise that we all involuntarily enable AI systems by using them and contributing to them. So the power of AI does not emanate from the technological features of a new kind of artefact, but rather relies on the capture and use of human capabilities within societal and media-technological arrangements designed for this purpose. AI systems are thus quite vulnerable to our availability as users and data producers.

However, the fact that AI systems are dependent on us is only one side of the coin. A technology that relies on deep engagement and near-constant interaction with billions of people can also lead to a serious backlash against those people. This backlash relies on the fact that we are all quite transparent to those systems through our dense interconnectedness with them that involves datafication of all areas of life. In this part of the book, I will look at this flip side of the coin, asking what kind of impact Human-Aided AI has on humans. Beyond the systemic diagnosis of serious power imbalance through capture and exploitation, I will argue that the essence of the power of data-driven AI is actually *prediction power* – that is, the capacity to predict unknown information or the future behaviour of any human. Heuristically, a form of 'intelligence' that emerges from analysing the behaviour, movements, thoughts, reactions and other data streams originating from all niches of the lives of many people is equivalent to the ability to classify, discriminate and predict the behaviour, movements, thoughts, reactions and so on of human beings. If data-driven AI presents an elaborate form of pattern recognition capability that is in real time calibrated on the data streams of millions of individuals or cases, this same capability can be used to predict, differentiate, sort and score the different individuals and cases represented in the training data stream.

I thus argue that the principle of prediction is central to, and even defines, AI in its current form, which has emerged since the mid-2010s. I henceforth use the umbrella term 'predictive analytics' to refer to AI systems that are explicitly built for predictive purposes, typically making use of machine learning to predict unknown information or the behaviour of individuals, or future developments of cases. Predictive analytics is one of the largest markets for AI applications today, ranging from targeted advertising to automated hiring, predictive policing and border control. Beyond the application of AI technology for explicitly predictive purposes, I maintain that the ability to predict is central to all major examples of AI technology, including generative AI models such as large language models that rely on the ability to predictively continue a list of words (tokens) by predicting the most probable next word (token) to thereby generate a sentence or paragraph of text.

In this part of the book I will argue that the predictive capacities of contemporary AI come with specific ethical and political challenges for which our critical debates are not yet well enough equipped. To this end I will look at the more elementary and basic examples of predictive AI, rather than the fancy examples that have been debated in the last months around generative AI and 'frontier AI'.[1] This serves a dual purpose. First, predictive capacities are at the core of all contemporary data-based AI systems and constitute the specific type of power of those systems, which is prediction power. Bringing the pervasiveness of this problem to light requires looking at mundane and widespread examples. Second, recent debate of AI risks and harms overly focuses on 'frontier AI', neglecting the already imminent risks of more classical or narrow machine learning applications, which have been out there for years, impacting the lives of millions, and are therefore deserving of more critical and ethical attention.

Predictive analytics: functional characterisation

To give a direct example of what predictive analytics could do, we can quote the well-known paper by Kosinski et al (2013) titled 'Private traits and attributes are predictable from digital records of human behaviour'. The authors showed empirically that a social media platform such as Facebook can, from a handful of behavioural data points (Facebook likes), derive a range of personal information about a platform user. Are you single or in a relationship? Did you grow up with separated parents? Do you smoke, abuse alcohol or other substances? What is your ethnicity or religious affiliation? What are your political views? Are you homosexual? What is your gender? Another paper, 'Evaluating the predictability of medical conditions from social media posts' by Merchant et al (2019), showed that detailed predictions about diseases like diabetes, hypertension or sexually transmitted infections can be derived from the content of social media postings.

These are just two examples from a huge corpus of research that highlights how behavioural data that are produced and recorded routinely from users of networked services can be used to predict personal or sensitive information that is normally hard to obtain. By the term 'behavioural data' I refer to the data that result from merely using networked media platforms, including social media sites and smartphone apps – that is, data about what content we view, like, share or bookmark. Facebook, the platform that was addressed in the study by Kosinski et al, is of course only one example of this, but the same applies to all other social media platforms. Predictive analytics also works with other usage data that can be collected outside of platforms, such as browsing history, tracking cookie data, location history or credit card transactions. Tracking data on the internet is used, for example, to assess insurance risks, creditworthiness, personal interests, purchasing behaviour, education level, substance abuse and addictions, demographic information, political views, gender, ethnicity and religious affiliations.[2]

By 'predictive analytics' I am broadly referring to the academic, scientific and industrial practice of researching, building and using predictive models. The term 'predictive model', as I use it, is not specific with regard to any particular kind of algorithms. These can comprise machine learning methods of various kinds, but also simpler regression analyses or statistical evaluations. For the ethical debate, we need a functional characterisation of this technology based on what inputs it requires, what outputs it provides and how it is integrated into real-world use scenarios. That is, an ethics of predictive analytics is not about opening the black boxes of predictive models, but about understanding their functional properties and integration in larger sociotechnical constellations.

From a functional perspective, a predictive model is first a 'machine' that receives as input the known data about an individual or case, henceforth referred to as 'auxiliary data' (for example, tracking data or social media likes), and outputs an estimate of unknown information about the individual or case, referred to as 'target information' (see Figure 5.1).[3] The key question is, of course: How is a predictive model able to derive meaningful predictions from the auxiliary data that are generally very different in kind from the target information? Instead of the rather complex and algorithm-specific way of answering this question, in terms of our functional characterisation of predictive analytics it is better to ask this question in another way: *Who* can actually make something like this? *Who* can make machines that use readily available data to meaningfully estimate hard-to-find or highly sensitive information regarding just about anybody?

This is typically something that only big-data companies can do – for example, a large social media platform with millions of users who leave data

Figure 5.1: Functioning of predictive analytics in social media

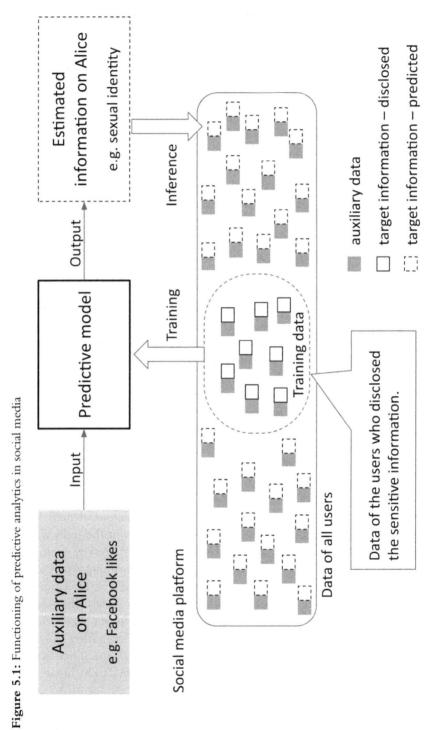

Source: Rainer Mühlhoff

traces every day. Let's take the company Meta, with its platforms Facebook and Instagram, as an example. In Figure 5.1, the grey boxes correspond to the collected usage data of each platform user. A few of the users not only leave behavioural data such as likes, posts or group memberships on the platform, but also disclose certain personal or sensitive information about themselves; for example, some users provide information about their sexual orientation in their platform profile. Such information is shown as white boxes with a solid border in Figure 5.1 (we're looking only at the solid, not the dashed white boxes for the moment). Whenever a grey and a solid white box are placed next to each other, this corresponds to a user who has disclosed *both* behavioural data (which everyone on the platform discloses) *and* personal information about their sexual orientation.

Let's assume that only a few per cent of the users of a social media platform are willing to give explicit information about their sexual orientation. In the case of Facebook, this still amounts to several hundred million users – in Figure 5.1 they are combined into the circled area in the middle. The idea of predictive analytics is to use the data provided by this subset of users to train, using supervised learning, a predictive model that outputs a prediction of the white data based on an input of the grey; in our example, this would be a predictive model for estimating sexual orientation based on Facebook likes, posts, group memberships and so on. That is, the data of those (fraction of all) users who voluntarily disclose the target attribute (sexual orientation) alongside the auxiliary data can be used as a training dataset for a machine learning model that automatically looks for correlations between the auxiliary data and the target information in this set.

Once the platform company has trained a predictive model for sexual orientation from the data of this minority of users, that model can be applied to *all other* users – that is, to all those who are only marked with a grey box in Figure 5.1. The dashed white boxes represent these subsequently predicted data about the sexual orientation of users who did not disclose this attribute explicitly. The model trained on the data of the 'permissive' minority thus allows the platform to estimate the sensitive target information even about those users who potentially do not wish to provide it. Hence, if you use a social media platform anonymously or specifically withhold certain personal attributes, you can assume that the platform is still able to ascertain this information about you. This is because machine learning technology allows the platform company to *leverage the data many other users have voluntarily provided* to predict 'missing' information about you.

The dual business model of digital media

The principle of predictive analytics forms not only the core of what contemporary machine learning does, but also one of the fundamental ways in which the contemporary digital economy operates. Platforms are extremely interested in building predictive models from usage data because these can be lucratively marketed: in the finance and insurance industries, in advertising, in human resource management and so on (O'Neil, 2016). The idea of leveraging available data – not about a specific individual, but about masses – to gain an informational advantage with regard to information unknown about some other individuals or cases forms a pervasive scheme of value generation and capital accumulation in what Shoshana Zuboff calls 'the age of surveillance capitalism' (Zuboff, 2019). The idea of monetising data through predictive analytics has its precursor in a well-known fact about the internet economy of the 1990s and 2000s: many web and software services started to be offered to users for free because they were being financed through advertising. This was the age of advertisement banners on websites that were the same for all visitors. Since that time, things have developed in two ways. First, advertisements have become personalised by means of models that predict what each user is most likely to click on. Second, it turned out that it was not only advertisers that were willing to pay for access to such predictive models; risk management in finance and insurance has always been another prediction-hungry customer, and all other applications mentioned in the previous pages followed suit. This development opened up many potential markets for 'prediction products' (Zuboff, 2019), which are sometimes hidden markets in that they are hardly visible from the consumers' point of view. After all, the user rarely sees the direct impact of, for example, prediction-based insurance, credit risk management or automated hiring decisions.

The general culture of monetising digital products through data has thus led to a landscape in which providers of free digital services usually work with dual-faced business models. One is a front-end and the other a back-end business model. At the front end, a product is offered that seems impressively practical and therefore spreads widely, maybe even going viral. At this front end, you're a user, not a customer. Only at the back end are the paying customers of these companies acquired – these paying customers are companies interested in predictions about the platform's users (see also Zuboff, 2019: Chapter 9). In this way, Google, Meta, Apple, Microsoft, Amazon and many others sell access to (predicted as well as recorded) information about their front-end users on the back-end market. This can generate more profit than direct monetisation of the front-end services – for example, via a usage fee.

Thus, data companies sell the output of predictive models in market segments that are different from their front-end products, such as credit scoring and insurance risk assessment, automation of selection processes for jobs or university places, human resource management and hiring, individual pricing for products of all kinds ('differential pricing'), personalised advertising, product recommendations, youth protection, policing, border control, the judiciary and so on.[4] The data industry typically does not sell the predictive models themselves, which remain proprietary, but rather sells access to them or comprehensive services that, for example, handle the placing of advertisements on behalf of the advertiser based on predictive models (see Ruschemeier, 2023c). In many back-end markets, major platform companies have implemented auction mechanisms for accessing the outputs of predictive models, so that the highest bidder can buy the predictions. Here, the key concern is only about maximising profits; moral considerations usually only play a role after the application of massive public pressure (Noble, 2018; Milmo, 2021; Wong, 2021). Ordinary users typically see nothing of these back-end markets, because this level of business activity by platform companies takes place just below the surface of the colourful front-end interfaces. In terms of scale, however, the back-end business activities of these companies should be viewed as the part of the iceberg beneath the surface: compared to what a plain, minimalist, publicly visible front-end interface suggests, the back end business activities are usually much more extensive, because this is where the company's revenue is generated.

Targeted advertising

Targeted advertising, often referred to as 'personalised ads', is, next to credit scoring, a model case of predictive analytics.[5] Today's targeted advertising is ineluctably tied to advertising in online environments, where it is possible to automatically select different ads for different visitors to the same location (visitors to a website or users of a smartphone app, for instance). Why is targeted advertising a use case of predictive modelling? Because the targeting decision for ads relies on an automated prediction of user behaviour. The targeting decision for a given advertisement means selecting, out of many available users of or visitors to a platform, website or app, those most likely to be interested in the ad. In personalised advertising, this targeting decision is delegated to an algorithm that predicts, for any combination of user and ad, the likelihood that the user will click on that ad.

As I have argued together with Theresa Willem in a joint work pointing out the specific ethical challenges of personalised advertising (Mühlhoff and Willem, 2023), we need to understand its novelty by delineating it from other targeting mechanisms.[6] As personalised advertising is only possible in online environments, a good starting point is to compare it to its traditional counterpart: offline advertising. A common form of offline advertising is street or magazine advertising. Advertisers choose a location, such as a specific street corner, shopping mall, city district, sports event, newspaper or magazine, for a fixed advertisement to be displayed to all visitors to that location over a certain period of time. While the location is carefully chosen to match a specified audience, typically every person who passes by sees that advertisement, and the same version of it. This form of audience targeting might thus be called 'static targeting' (Mühlhoff and Willem, 2023) because there is no individualised (per visitor) and/or responsive decision mechanism that enables the ad to be shown only to a selection of people walking by. In theory, static targeting is also possible online, such as when a fixed ad is placed on a website so that it is equally visible to all visitors. However, static targeting is weak in comparison to more elaborate contemporary online targeting methods and does not play an important role in today's online advertising, as it represents the technological state of the art in the pre–social networking era of the internet.

The technological development of social media enabled a profound revolution in advertising due to its ability to aggregate extensive, high-resolution individual data profiles (see, for example, van Dijck, 2009). This came with three major qualitative leaps in ad targeting.

The first leap is towards a methodology that might be called 'explicit targeting' (Mühlhoff and Willem, 2023): online advertising can base the targeting decision on a per-user matching of targeting criteria against explicitly recorded personal attributes. For example, on social media platforms, users sign up by providing a personal profile, as part of which they voluntarily disclose personal information, including data on their age, gender, relationship status, personal interests, group memberships, friends, educational background, work affiliations and so on. Providers of explicit targeting are mostly large platform companies because they need a broad base of users who are incentivised (through mechanisms often unrelated to advertising) to provide as much personal information to the system as possible. On many platforms and app-based services, users are motivated (or sometimes obliged) to submit personal information to the platform when they communicate with other users (likes, photos, messages and so on). Monetising this information through advertising is the main source of revenue for many social media platforms (Statista, 2022).

Compared to static targeting, explicit targeting frequently renders the specifying of a *location* (such as a specific website on which the ad is shown) obsolete. The location, to begin with, was always only ever a proxy for personal attributes, which can be specified in this new form of targeting. Predetermined by the individualised targeting criteria, the platform automatically 'decides' at which location (that is, at which point in the interaction between user and online service) which ad is to be shown.

The second qualitative leap comes with individual targeting mechanisms that rely not only on personal attributes explicitly specified by the users of a platform, but also on predicted information. Imagine a user did not specify their age or gender in their profile, while their extensive behavioural data (posts, likes, browsing history or social graphs) are available to the platform.

This is precisely the situation where predictive modelling comes in: using machine learning and data analytics techniques, the available behavioural and tracking data can be leveraged to estimate relevant targeting attributes about individual users in cases where such attributes are not explicitly provided.

This method, however, is not limited to 'filling the gaps' in users' profile information. It also enables a covert expansion of the available targeting parameters to include information that hardly any user would voluntarily provide. This information includes, for example, purchasing power, wealth or credit risk scores, as well as estimated information on medical conditions, sexual orientation or ethnic or racial background (O'Neil, 2016; see Mühlhoff, 2021). As Andreou et al (2018) explain, *any* data that users share with the platform are treated as 'raw data' input that can potentially be 'translated' by data analytics algorithms into 'targeting attributes [...] that advertisers can specify to select different groups of users' (Andreou et al, 2018).

It is a reasonable assumption that most contemporary online advertising providers combine explicit and predictive targeting without a clear separation of the two. Advertising customers merely specify certain targeting criteria that will be matched by the social media platform either with data explicitly provided by users (where available; for example, for users who disclose their age in their profile) or with predictions of the same parameters when a user did not provide that information. This way, a social media platform like Facebook is able to offer to its advertising customers an extensive list of hundreds of fine-grained targeting criteria. Figure 5.2 includes screenshots from the Facebook back end for advertising customers, showing a small sample of the extensive catalogue they can choose from.

The third leap is predictive targeting, as an 'improved' version of explicit targeting, which makes estimated audience attributes available as targeting criteria. This next qualitative leap in the development of targeting methods presents a way to do it altogether without the advertiser specifying targeting criteria. One version of that approach is evident in what we termed 'lookalike targeting' in Mühlhoff and Willem (2023). Here, advertisers provide a list of relevant users who are known to respond positively to the campaign. A targeting algorithm is then used to find 'similar' users on the platform (with the criteria for similarity being determined 'algorithmically'), who are then shown the advertisement.[7] This mechanism entails that the targeting algorithm automatically determines behavioural and demographic markers that are specific to the sample group and then applies this analysis to all other platform users to show the advertisement to those who are deemed to be similar (Andreou et al, 2019; Semerádová and Weinlich, 2019).

The power of this targeting principle becomes evident specifically in the context of social media – for example, in combination with Facebook groups, fan pages or the collection of email addresses from offline events.

Figure 5.2: Screenshots from the Facebook ad-targeting back end in 2021, showing available targeting criteria

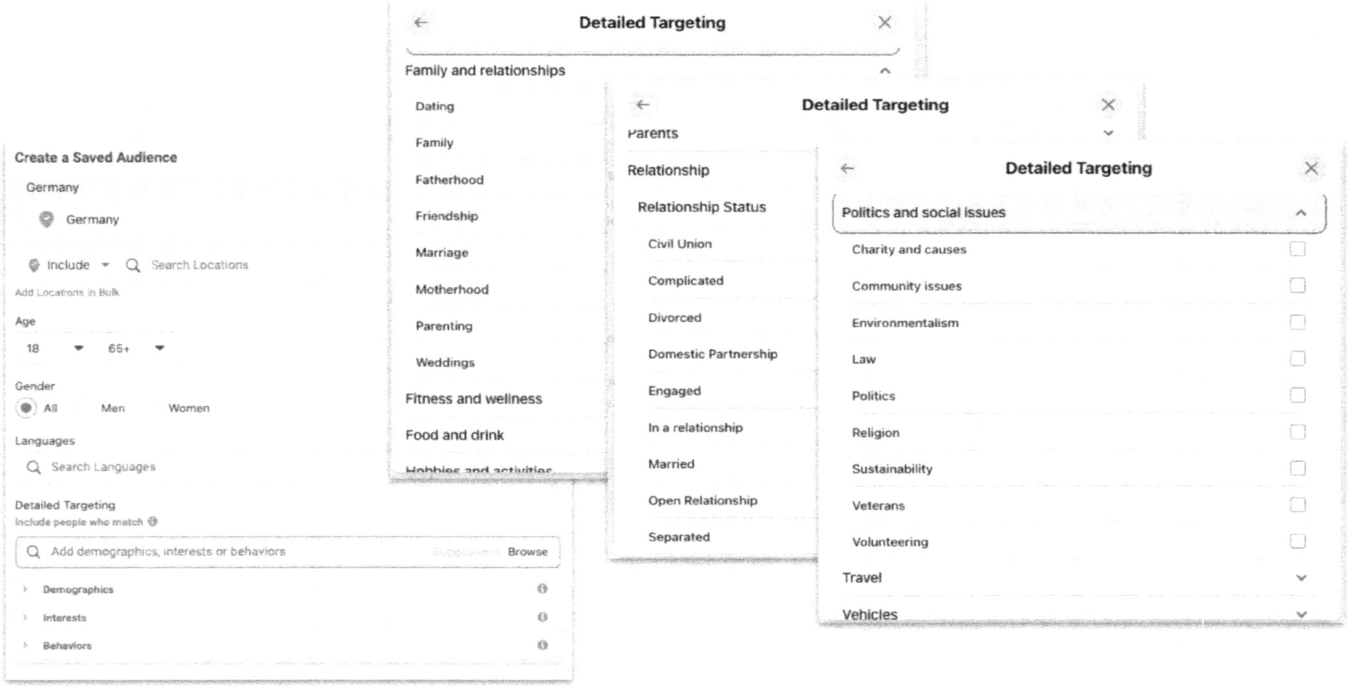

Source: Rainer Mühlhoff, screenshots from facebook.com, August 2021

Facebook groups are online interest groups to which users subscribe in order to engage with other users about a shared interest, such as arts and entertainment, products, leisure activities, political views or a shared disease or medical condition. Lookalike audiences on Facebook can be created from the members of such groups, as well as from the followers of political organisations, activist groups, religious associations or minority movements. Alternatively, advertisers can upload lists of email addresses of known customers from which Facebook creates a lookalike audience. This targeting type is known to have been mobilised in Donald Trump's 2016 presidential campaign (see Baldwin-Philippi, 2017).

Case study: psychological targeting in election campaigns

Since the Facebook/Cambridge Analytica scandal in the wake of Donald Trump's election in 2016 (Grassegger and Krogerus, 2016; Rosenberg et al, 2018), the public has become aware of the principle of predictive targeting of election advertising. The now insolvent data analytics and political consulting firm Cambridge Analytica used Facebook data to create predictive models to estimate psychological character traits of users, which were then used to influence elections through microtargeting of political ads tailored to the individual user's presumed psychological dispositions and political views. Cambridge Analytica's former CEO Alexander Nix claims his firm was involved in more than 40 election campaigns in the US alone, including the campaigns of Ted Cruz and later Donald Trump in the 2016 presidential election. Cambridge Analytica was also linked to the Leave. EU campaign in the UK's Brexit referendum that same year (Sellers, 2015; Confessore and Hakim, 2017; Hern, 2018) and has been active, sometimes through its parent company SCL Elections, in many countries of the Global South (Ghoshal, 2018). One of the datasets that was used by Cambridge Analytica to train their predictive models was data on about 80 million Facebook users collected by Aleksandr Kogan, then a psychologist at Cambridge University, as part of 'academic' research on Facebook users (Grassegger and Krogerus, 2016; Rosenberg et al, 2018). Kogan was working for Global Science Research, a company he founded, which collected these data using a Facebook app – in other words, software from the company embedded in Facebook – which took the form of a kind of psychological quiz and which the company later sold to SCL, Cambridge Analytica's parent company.

Which models Cambridge Analytica was really able to train on the basis of such data has never been conclusively proven. A plausible version of the story, disseminated also by Alexander Nix, relates that the predictive models made an assessment of psychological character traits according to the OCEAN model of psychological trait theory (often called the 'Big Five personality traits' or 'five-factor model'). This model classifies people along the five dimensions of openness, conscientiousness, extraversion, agreeableness and neuroticism (Matthews et al, 2003: 23–4). It is plausible that Cambridge Analytica was able to train predictive models that allowed them to obtain estimates of these character metrics from the Facebook usage data of *any* Facebook user. Such predictive models could presumably be trained using the psychological data of 80 million Facebook users acquired from Kogan's research in combination with platform usage data of these individuals (grey and white data points in Figure 5.1). Using such a predictive model, Cambridge Analytica was then able to estimate a detailed psychometric profile across *all* Facebook users, not only those whose data were leaked.

Once one has created such a predictive model for psychological character traits, the question follows as to how it can be used to influence elections. In relation to the campaign of Ted Cruz in the primaries of the 2016 US presidential election, it became known that Cambridge Analytica had designed advertising messages specifically tailored to different psychological personality types (see Figure 5.3) (Davies, 2015; Confessore and Hakim, 2017). A predictive model that uses Facebook usage data to group users into the relevant psychological groupings could then be used to automatically display to each user the version of advertising material that best 'fits' their personality profile. Alexander Nix explained in a presentation about the Ted Cruz campaign that the goal was to mobilise voters who were in favour of preserving the Second Amendment (the right to bear arms) (Nix, 2016). To this end, individuals classified as 'neurotic' and 'conscientious' were shown the advertising message seen on the left in Figure 5.3, with the slogan 'The Second Amendment isn't just a right. It's an insurance policy. / DEFEND THE RIGHT TO BEAR ARMS' displayed on top of a close-up photograph of a broken front-door window featuring a black-gloved hand reaching for

Figure 5.3: Two advertising messages from the Ted Cruz campaign presented by Cambridge Analytica CEO Alexander Nix in a lecture at the Concordia Annual Summit, in 2016

Source: Cambridge Analytica (2016), from https://www.youtube.com/watch?v=n8Dd5aVX LCc 4m40s

the door handle from outside. Individuals classified as 'agreeable' and not so 'open' were shown the version shown on the right, with the slogan 'From father to son / Since the birth of our nation / DEFEND THE SECOND AMENDMENT' in front of a backlit photograph at sunrise, showing the silhouette of an adult and a child in a meadow landscape, carrying weapons pointed at the sky and the adult apparently pointing at passing birds.

Such an approach is called 'psychological targeting', because the target group of an advertising campaign is subdivided according to (estimated) psychological characteristics, and subgroups are addressed with adapted messages. This procedure falls into the more general category of microtargeting, a communications strategy that, in the context of elections, identifies (micro)groups that are precisely defined as possible targets and whose members' behaviour can be influenced with tailored messages (for example, undecided voters in swing states). By defining the smallest possible groups for tailored messaging, resources for advertising can be used very efficiently, as the number of people seeing a message that may have little effect on them is reduced. The procedure allows different targeting groups to be provided with different messages in the manner described earlier to minimise the risk of a public backlash against particularly polarising advertising messages. The fact that microtargeting has been used in the context of US elections is not a new phenomenon. The Barack Obama campaigns in 2008 and 2012 already operated in this way – it can even be argued that they were pioneers of these techniques, insofar as microtargeting first became known worldwide through this type of campaigning (Tufekci, 2014).

So what makes the Facebook/Cambridge Analytica scandal so dramatic? It reveals a new quality of *collective impact*. This quality lies in the fact that the data of some Facebook users could be used to train predictive models for psychological characteristics that could subsequently be applied to *all* US Facebook users in order to group them according to psychological characteristics and provide some of them with tailored advertising. So, in contrast to how the scandal was publicly debated, the main provocation was *not* that the data of Kogan's 80 million Facebook users were leaked. The crucial point is that these data from 80 million users, who volunteered to have their character traits measured in a psychological quiz, were used to predict psychometric information about millions of *other* individuals who knew nothing about it and who had not themselves taken the quiz. The majority of the damage resulted from the fact that, by means of a predictive model, Cambridge Analytica was able to manipulate the voting behaviour of *any* Facebook user through tailored political ads. The damage that was done to the 80 million quiz participants, inasmuch as their (anonymised) quiz data were abused, is only minor by comparison.

The Facebook/Cambridge Analytica scandal sparked many debates on the potentially negative influence of social media platforms and their

AI-based data analytics capabilities on democracy. Importantly, the problem is actually much bigger than the Facebook/Cambridge Analytica scandal; after all, Cambridge Analytica does not exist any more. The phenomenon, however, does not hinge on the shady activities of third-party app providers like Cambridge Analytica, as the platform companies (in particular Meta, Google, Amazon, Microsoft and Apple) *themselves* have all the relevant data to train predictive models for the microtargeting of advertising. To stay with the Facebook social media platform for a moment: as mentioned before, Figure 5.2 shows Facebook's catalogue of targeting criteria for advertising. The list includes numerous data fields that most platform users would not provide about themselves. In such cases, Facebook estimates this information using predictive models. Hence, it doesn't need an outside company like Cambridge Analytica for manipulative data use of this kind to be offered as a service on the advertising market. The real scandal about the Facebook/Cambridge Analytica affair is *not* that an outside company obtained the data of 80 million Facebook users. Rather, the core problem is that whoever has aggregated social media data – first and foremost, the platform companies themselves – can train detailed predictive models with those data, and apply these models to all other users on the same platform. Google, Apple, Facebook, Microsoft and Amazon are all active in the business of microtargeting and individualised advertising, thus leveraging the rich usage data they routinely collect from their users.

Prediction power

The preceding discussion shows that data accumulation gives rise to a specific form of *power* if it is used for predictive analytics. Only those who possess data can train predictive models, and these models come with the ability to create predictions about *any* individual. Hence, these models instantiate a certain general capacity that I refer to as 'prediction power' (Mühlhoff and Ruschemeier, 2022, 2024a; Mühlhoff, 2023b, 2024). Prediction power is a key aspect of the power of platform companies. Due to the economic significance of predictive AI applications as sources of revenue for platform companies, prediction power currently represents one of the most important forms of informational power asymmetry between data-processing actors and individuals (Mühlhoff and Ruschemeier, 2024a; Ruschemeier, 2024).

By conceptualising the issue as a phenomenon of power, a social theory angle enters into the critical debate of predictive analytics. Possession of predictive capacities makes actors powerful in the sense that, by using and abusing their predictive capacities, they could shape diverse social, political, economic and cultural domains. The framing of predictive technologies as prediction power is key to the debate around regulatory proposals that aim to control and mitigate prediction power (see Chapter 6). As we shall see, what needs to be regulated is the *potential* – that is, the latent capability – for certain (harmful) actions (see also Mühlhoff and Ruschemeier, 2022: 43). The mere possession of a predictive model in the wrong hands poses a widespread potential for harm, including discrimination, manipulation through the exploitation of vulnerabilities, influence on democratic discourses and so on, that could affect a multitude of unspecified individuals. This latent general danger, which comes with a predictive model that could be applied to anybody, constitutes the power position of those who hold predictive capacities.

Putting election campaigns aside, isn't personalised advertising harmless in most cases?

Personalised advertising can be manipulative, especially when it exploits predictions about users' psychological or emotional dispositions. In the case of election advertising, this kind of targeted influence can be decisive for the outcome of an election, especially if it is used in contexts where a few votes might make a big difference (for example, tipping-point combinations, swing states).

But how would an ethical assessment of microtargeting turn out if we could rule out the *manipulative* use of predictive models? Let's assume that microtargeting for election advertising was prohibited, and also targeting according to (predicted) psychological and emotional criteria. Would personalised advertising, such as for products or events, still be ethically questionable? A current advertising campaign by Facebook itself aims to convince users of the supposed added value of personalised advertising. It puts forward the old argument that personalised advertising allows users to see more advertising that is 'relevant' and 'interesting' to them and promises them that they will be less disturbed by generic ads (see Hutchinson, 2022 for an example). Wouldn't this advantage actually outweigh the ethical concerns? Wouldn't targeted advertising – within the aforementioned hypothetical limits – be one of the most innocuous applications of predictive modelling?

The answer is no, *not at all*. And I would like to use another example to make this case.[8]

One area of application in which personalised advertising is due to have a major impact over the next few years is research into new drugs and therapies in clinical trials. These are large-scale studies with test subjects by which new drugs or therapies are tested. Imagine you want to test a new therapy for treating type 2 diabetes, and for that purpose you need thousands of participants with this disease.[9] Instead of placing ads for your trial on public transport, in doctors' offices or in illustrated magazines, for example, you place ads on social media platforms using demographic targeting criteria to narrow down the target group more efficiently. Proponents of this social media–based recruiting strategy for clinical trials argue that, this way, you can reach a more relevant population for the same advertising budget than with broadly distributed offline advertising, and in the case of medical research, this cost efficiency would ultimately benefit us all.[10]

To explain why this kind of advertising is such a problem from a data ethics and data protection perspective, let's briefly review the chain of relevant data-processing steps involved in this kind of advertising. First, the advertising customer provides the social media platform with the ad and some (initial) targeting criteria (for example, demographic attributes). Second, the platform's targeting algorithm, a predictive model, determines which individuals would most likely click on the ad, and the ad is displayed to them. However, digital advertising providers, including Facebook and Google, do not only display personalised ads to users; rather, third, they also record users' *engagement* with the ads. For example, they register whether the user looks at the ad (how long they pause there while scrolling) and whether they click on the ad or maybe even 'share' it (that is, send it to someone else on the platform). If the ad has an engagement opportunity such as the 'Sign Up' button,[11] this kind of direct engagement will also be registered by the platform.

Now, what happens with the engagement data collected in the third step? These data are not only used for per-click invoicing of the advertising customer. Much more importantly, these data can be fed back as training data into the predictive model responsible for the targeting decision. Engagement data are real-time verification data for the targeting algorithm as they yield an ex post validation regarding which of the predicted users really clicked on the ad and which did not. This, fourth, can be used to retrain and thus refine the predictive algorithm to more accurately predict which users are likely to click on that specific ad in future. In this way, the predictive ad-targeting system becomes more accurate over time, even exceeding its accuracy compared to the initial targeting criteria provided by the advertising customer in the first step. It is precisely this effect that has been described in scientific publications.

For example, scientists in the US wanted to use Facebook ads to gather cannabis consumers for a study on cannabis consumption behaviour (Borodovsky et al, 2018). In their study, they describe how the targeting of this ad was relatively poor in the beginning (because it is difficult to narrow down cannabis consumers by openly available targeting criteria such as demographic attributes), but showed a steep 'learning curve' in the first few days so that the targeting became notably more efficient after a few days (p 2). This indicates that the algorithm is constantly learning from ad-engagement data to improve its predictions. While very few users explicitly state on Facebook that they consume cannabis, the predictive model is able to interpolate this unknown information after a short time from how users react (or don't react) to the ad. The same holds for the targeting algorithm for patients with type 2 diabetes.

So, while an ad campaign is running on a social media platform, based on the users clicking or not clicking on this ad, a predictive model can be trained in real time that learns to predict more and more accurately from platform usage data those people to whom the content of that ad is 'relevant'. This predictive model belongs to the social media platform because it is created by means of proprietary data from that company. What is this model in the case of our type 2 diabetes advertising? It's a model that correlates with medical information. It's a model that can predict whether any Facebook user is likely to have type 2 diabetes. By engaging or not engaging with the ad, users unwittingly reveal highly sensitive information that is fed into the predictive model as training data. Using this predictive model, the platform can predict about *any other* user, including future users and users who never see the ad, whether they suffer from a similar condition. With the implicit support of the external research organisation that has commissioned the advertisement for a clinical trial (for medical research, these are often public institutions), the social media platform is enabled to estimate medical information about *any* Facebook user, today or in the future (Mühlhoff and Willem, 2023).

This scheme exemplifies a comprehensive and serious, but at the same time widely neglected, data ethics and data protection problem connected to personalised advertising. It must be assumed that the providers of digital advertising train a predictive model for any topic on which advertising is placed, which can then be applied to any third-party user in order to estimate that user's relationship to the content of the advertising. Imagine the topics in the medical field alone, where this is a highly explosive development. Think of advertisements for alcohol and drug withdrawal therapy, HIV prevention, hepatitis B therapy, therapy for psychological conditions such as depression, anxiety disorders or paranoia. But equally, consider examples outside the medical field, such as advertising for diets and cosmetics, financial products and loans, or advertising related to political or religious viewpoints.

It is important to remember that the engagement data and the predictive models trained with them remain in the hands of the platforms. These data, and the targeting models trained from them, are in fact the platform's main asset. It is safe to assume that these targeting models will be reused in other ad campaigns and repurposed as predictive models in other industry sectors – such as for risk scoring in the insurance industry, for credit scoring or for automated selection of job applicants.[12] In these sectors, there are financially strong customers for such prediction products devised from the use of engagement data. Of course, the customers from these sectors are not visible at the front end of the platform, but they are served at the back-end business operations, so we never see them as 'ordinary' users.

This shows that predictive targeting of advertisements is another instance of a Janus-faced business model on the part of the platforms. Through running the ad campaign, for which the platform gets paid by the advertising customer, the platform is given the opportunity to collect data that can be even further monetised in prediction-based products sold on other markets. In many conceivable cases, second use of the data and models resulting from ad campaigns comes with high societal risks and potential harms to already vulnerable groups. This falsifies one of the most pervasive arguments commonly advanced in favour of personalised advertising in relation to clinical trials: especially in the case of rare or stigmatised diseases, where test persons are typically hard to reach, proponents argue that with personalised advertising, more relevant candidates can be found. While this might be true at first sight, unregulated use of anonymised data and predictive modelling can easily turn this initial advantage against those already vulnerable populations, as the latter might face further discrimination in other sectors when predictive models are reused that can be trained on feedback data from ads.

6

Predictive Privacy

From the previous discussion of key applications of predictive modelling, we see that this technology comes with a specific set of challenges that might be phrased as privacy and data protection issues. Following this lead, I will now argue that these data protection challenges comprise two structural dimensions. First, predictive analytics is characterised by an unexpected *escalation of data sensitivity*: that is, from readily available data about an individual (for example, Facebook likes), information that is usually difficult to access (for example, the likelihood of substance abuse) can be estimated using predictive models. Second, predictive analytics has a *structure of collective enabling*: the data some users provide voluntarily to the platform facilitate the training of predictive models that can then be applied to any *other* users to estimate information about them. This chapter will discuss the implications for data ethics and data protection of these two structural properties of predictive analytics.

A new form of privacy violation

It is peculiar to predictive analytics that this technique allows an estimation of 'sensitive' information based on data that most users consider relatively innocuous, such as their social media usage or browsing history. Predictive analytics thus enables an *escalation of the sensitivity* of the data. The user has 'only' posted some likes on a social media platform, shared their GPS location or purchased some items on Amazon, but in reality the platforms can derive from these data estimates of much more sensitive information, such as the users' likely sexual orientation, whether they abuse substances or whether they are mentally unstable or pregnant. Such predictions could potentially include information that the user themself doesn't know – think of disease prognoses or credit scores.

Along with this first structural feature, then, we must note that predictive analytics potentially invades the privacy of the affected person – that is, of the person *to whom the model is applied*. Notice that there are two different groups of data subjects involved in predictive analytics: the training data subjects and the target subjects, who are the subjects to whom the trained model is applied (Khan and Hanna, 2023; Mühlhoff and Ruschemeier, 2024a, 2024b). It is the target subjects whose rights – in this case, to know and control what information about them is being processed by another entity – are potentially violated. A potential breach of the target subjects' privacy occurs here because personal information is *predicted* about the target persons without their knowledge or against their will (see in detail Mühlhoff, 2021). I argue that predictive analytics facilitates at large scale a new form of potential privacy infringement because it extends the range of common attack scenarios against someone's privacy to also include predictions. To see more clearly how this infringement constitutes a novel form of privacy attack, it is useful to distinguish the following list of three schematic types of attack (see also Table 6.1).[1]

Intrusive privacy violation: In the intrusive type of privacy violation, the attacker obtains unauthorised access to information that has previously been classified, protected or declared as private. This violation could happen, for instance, by breaking through access barriers, or by a trusted party misappropriating that information and passing it on without authorisation. Typical examples are hackers or spies who break through security barriers or exploit vulnerabilities in software or in organisational structures to gain unauthorised access to data. Other examples include fraudulent business practices, such as data brokers who obtain and resell personal data outside the scope of where their legitimate processing has been authorised by the data subject. The key qualitative aspect of an intrusive privacy violation is that it is a targeted attack on specified individuals by which 'actual' (as opposed to predicted) information is obtained. Intrusive privacy breaches are not limited to the domain of electronic data processing. For instance, the modern notion of 'privacy' emerged in US American discourse when the development and broad availability of photography posed new challenges to the privacy of celebrities and public figures (Warren and Brandeis, 1890).

De-anonymisation attacks: A completely different type of privacy attack became prolific in the 1980s and 1990s in relation to the statistical use of large personal datasets. For instance, records of medical treatment data hosted by hospitals or insurance providers were made available in anonymised form for statistical and research purposes. The strategy of a de-anonymisation attack is a way of linking records in such *deliberately published* but anonymised datasets to the original data subjects, thus breaking the anonymisation. Such attacks often leverage statistical properties of the published anonymised datasets and openly available background information such as demographic data. This makes them fully reliant on publicly available data without the perpetrator committing any form of intrusion, like breaking through or circumventing security barriers.

In a prominent case, in the US state of Massachusetts in the 1990s, the medical treatment data of around 135,000 state employees and their family members were 'anonymised' and compiled in a database for research purposes. A student at the Massachusetts Institute of Technology at the time, Latanya Sweeney, managed to reconstruct the medical records of the then Governor of Massachusetts, William Weld, from this dataset by lifting information from the Massachusetts voter register (Sweeney, 1997). This incident caused an uproar and greatly influenced the debate on data privacy in the US. In mathematics and computer science, Sweeney's attack fuelled research on database security and anonymity, with new, spectacular reidentification attacks on anonymised datasets being published regularly (Narayanan and Shmatikov, 2008; de Montjoye et al, 2013; Gymrek et al, 2013; Shokri et al, 2017). As a consequence, the term 'anonymous' is no longer the same as 'safe'; instead, it represents a criterion heavily dependent on context (Sweeney, 2002).

Sweeney's discovery also had a significant regulatory impact as it brought to light a new attack vector on privacy in the context of electronic mass data processing, which received considerable attention in the subsequent period, particularly in the context of medical data (this notably influenced the US Health Insurance Portability and Accountability Act; see also Ohm, 2010). Conceptually, the type of attack known as de-anonymisation is characterised as follows: like intrusion, it is a targeted attack on individual data subjects whose data are featured in the published data. However, no technical or moral access barriers are breached; rather, the basis for the attack is data that have been intentionally published with a promise of anonymity.

Predictive privacy violation: While intrusive privacy violations obtain unauthorised access to information, and de-anonymisation attacks reidentify individuals in anonymously published (statistical) datasets, predictive attacks on privacy present yet another qualitatively new attack vector, mainly for two reasons. First, information is obtained through making predictions instead of obtaining recorded information. Predictions are usually based on available but less sensitive data about an individual or case. Second, as elaborated in Chapter 5, in a typical scenario that involves machine learning models, the attack is enabled after many *other* individuals have disclosed sensitive information so that a predictive model could be trained using their data as training data. Hence, a predictive privacy violation is an attack that not only relates to one's own data but depends on collective data-sharing practices. According to this understanding, an individual's privacy is not only vulnerable when information they have *disclosed* somewhere (for example, to their doctor or in an online store) is misappropriated by the data processor or stolen by hackers. Rather, their privacy can also be infringed through information that is *derived* or *estimated* by leveraging the data many *other* people have voluntarily shared with a data or platform company.

Table 6.1: Qualitative comparison between the three attack types on privacy

	Intrusion	De-anonymisation	Prediction
Most relevant since …	1960	1990	2010
Violent breakthrough of access barriers?	Yes	No	No
Target subject must be in the data?	Yes	Yes	No
Only brings forth information explicitly captured about the data subject?	Yes	Yes	No
Based on mass data?	No	Yes	Yes
Targeted/individualised or wide-ranging attack?	Targeted	Targeted	Wide-ranging

Source: Amended from Mühlhoff and Ruschemeier (2022: 52)

This qualitative distinction between different types of privacy violation suggests that we need to refine our conception of privacy in the context of big data and AI technology. In fact, the conception of privacy has always been co-developing with the emergence of new technologies that pose specific privacy threats.[2] Predicted information, however, has so far not been typically included as part of the sphere of personal information in data protection and privacy contexts.

The concept of predictive privacy (first definition)

In order to debate the effects of predictive analytics as a privacy issue, we have to adopt an extended understanding of the ethical and political concept of privacy that includes *predicted* information as part of the private sphere. To give this extended understanding of privacy a name, I have been proposing 'predictive privacy'[3] and defining it in the following way:

Definition 1 (predictive privacy violation): The predictive privacy of an individual or group is violated when personal information is predicted about them without their knowledge or against their will, in such a way that it could result in unequal treatment of an individual or group (Mühlhoff, 2023b: 4).

The concept of predictive privacy essentially extends the domain of informational self-determination of a person P to include control not only over how factual information about P circulates, but also over information that is estimated about P by any third party. For this extended understanding of the sphere of personal information – the sphere of predictive privacy – to be potentially violated, it is irrelevant whether the predicted information is actually correct, because even incorrect predictions suggest that the individual is being judged based on a certain piece of personal information, which can have adverse effects on the person concerned.[4] The only requirement for an actual predictive privacy violation is that the prediction be made without the knowledge and against the will of the data subject – just as the disclosure of non-predictive information is a violation of privacy only if it has not been authorised by the data subject. As an additional condition, this definition assumes that in order to commit a predictive privacy violation, some differential or unequal treatment must arise as a consequence of the predicted information. While there are good reasons to skip this extra provision in the definition of predictive privacy, by introducing it we obtain a stronger (narrower) definition that is better suited as a norm in practical contexts. It spares us the idle debate about whether it is already a predictive privacy violation if someone privately and without consequences guesses personal characteristics about another person, which admittedly most people do on a daily basis.

A major challenge related to predictive privacy is how to create a general public awareness of the risk of privacy invasion through prediction. This entails that a comprehensive ethical and political debate is needed as to the implications of using estimated information in treating people differently. Debate is also needed to clarify how the ethical value of predictive privacy relates to other values against which it may need to be weighed in specific situations. For example, there are good reasons for patients in medical contexts to opt for the use of medical diagnostics based on predictive AI. In such a case, the patient could decide to forgo the preservation of their predictive privacy in favour of a potentially more effective medical treatment. A less trivial and more controversial matter, however, is the balancing of the predictive privacy of some individuals against collective values such as public safety. Is it ethically viable for algorithmically selected people to be pre-emptively searched or detained based on probabilistic estimates about their behaviour? Under the term 'predictive policing', algorithmic risk modelling has led to pre-emptive police operations, subjecting algorithmically selected individuals, geographical or demographic groups and neighbourhoods to increased police surveillance based on predictions of future criminal activity. In many US states, the criminal justice system uses predictive models of criminal recidivism probabilities (that is, of the likelihood that an offender will commit another crime in the future), such as when making decisions about parole or suspended sentences. Both in the police and in the criminal justice systems, such predictive models have been found to exhibit serious racialising biases and to perpetuate existing patterns of discrimination in the guise of a data-based and therefore supposedly objective assessment.[5]

Even aside from the issue of potential biases, it is dubious from an ethical point of view that we are allowed to use predictive modelling as the basis for decisions about security controls or sentencing. The concept of predictive privacy constitutes an ethical and political value that is violated in these examples irrespective of the (related but distinct) issue of bias. The ethical question that needs scrutiny is whether restrictions and unequal treatment are to be imposed on some people based on the merely probabilistic projections of their future behaviour – behaviour that they themselves can influence through the exercise of free will.[6] This issue will be extensively discussed in Chapter 7.

In this section I introduced the concept of predictive privacy in relation to the affected individual. From the perspective of the person to whom a predictive model is applied, predictive privacy violation is marked by an *escalation of sensitivity* of the information available to the operator of the system. From auxiliary data such as usage data or behavioural data that many people do not consider overly sensitive and so inadvertently disclose to service operators, more sensitive personal information about an individual can be estimated. This escalation of sensitivity is, as stated at the beginning of this section, the first of two structural observations about the challenges posed by predictive analytics.

Collective harms: our data affect others

The second structural feature of predictive analytics is even more explosive, and at the same time more counter-intuitive. When a predictive model estimates sensitive information about an individual P, it does so using the data from many *other* individuals (see the functional explanation of predictive modelling in Chapter 5). Hence, it is the data of many others that enable the violation of P's privacy. Predictive models are trained on the – possibly anonymised – data of many thousands to millions of people and are therefore only a by-product of our *collective* data-sharing practices. The individual P to whom the predictive model is applied need not themselves be included in the training dataset used to create the model. This means that all of us, when we use networked digital media, are strengthening the ability of platform companies to differentially treat and potentially discriminate against *other* individuals – and not necessarily ourselves.

This is an important message, especially to all those who think that they 'have nothing to hide' and can therefore leave their data to the ministrations of Google or Dropbox, Meta or Apple, Uber or Tinder, Strava or Flo. The data from the numerous supposedly 'normal' users (or users who consider themselves normal) make it possible to train predictive models that may then classify any other person as possibly deviant, unhealthy or insufficiently creditworthy, as a risk or as potentially dangerous. Our individual data permissiveness is therefore a behaviour relevant to the community; this also holds when we do not fear any negative consequences for ourselves or think 'the platform company already knows everything about me anyway'. In the current situation, where prediction technology is not properly regulated by law, *our* data enable companies to legally derive and place automated bets on the future behaviour or unknown personal attributes of *any other* person in our societies.[7]

Because predictive analytics leverages available mass data against any person – even those whose details and attributes do not themselves feature as part of the mass dataset – predictive analytics is based on a *structure of collective enabling*. This is the second characteristic feature of predictive analytics. These AI systems are only possible because many of us as users serially make the same decision to disclose data on a daily basis in ways that we subjectively consider as more or less harmless (this includes situations where we disclose data in anonymised form). The privacy, autonomy and equal treatment of arbitrary individuals thus become vulnerable to predictive analytics precisely because, for a sufficient number of *other* people, the negative effects of their data practices are not sufficiently noticeable or reflected upon. Often a minority of users implicitly or explicitly 'opting in', and thereby consenting to the collection of their data, is sufficient to train predictive models that can be applied to anyone else, such that this minority is effectively and unwittingly setting the baseline of privacy that holds also for all other users.

One way of imagining the collective aspect of one's data behaviour is by the analogy to environmental pollution from car exhausts. Here, too, the effects of our actions – the emissions we create – affect many others, and not even primarily ourselves. Moreover, each individual contribution, considered in isolation, is marginal. But it is the combination of the behaviour of many that leads to a catastrophic overall effect. Analogously, we can think of the data generated by our daily use of networked digital services as 'data pollution' (Ben-Shahar, 2019). Economically speaking, the data we produce come with social externalities in much the same way as atmospheric emissions are an externality. These externalities are costs induced by our data practices, but they do not necessarily accrue to ourselves and thus are not priced into our individual cost-benefit assessment upon making our choice to use certain digital services, but must be covered by others or the community.

The analogy to environmental pollution is a handy metaphor and is useful in terms of illuminating the risks, for instance, as part of public awareness campaigns. As a philosophical and scientific concept, however, the analogy has its limitations. First, while it is indisputable that exhaust gases represent purely *negative* externalities, data aggregation can have both *positive* and *negative* effects. After all, there are applications of predictive models that are useful to individuals and society, such as improving medical diagnostics or recommending interesting content to media users. In thematising data as a social externality, there is the core ethical problem of having to distinguish between socially desirable and undesirable applications of predictive analytics in order to decide whether the sign of the externality is positive or negative. Second, the analogy breaks down in terms of how much the externality scales with the number of co-polluters. Between atmospheric exhausts and air quality there is a roughly inverse-proportional relation – that is, if only half as many people drive cars, the air quality will be around twice as good. However, the same is not true with respect to social media data collection: half as many users providing their data are often still enough to train a predictive model that will subsequently be applied to everyone. And even if accuracy decreases in such a situation (which it likely does), the models are trained and used anyway.

Predictive privacy as a collective interest (second definition)

Even to communities that are savvier when it comes to issues around data protection and privacy, the collective enabling structure of predictive analytics is a new and sometimes hard-to-grasp idea. Privacy is often viewed through an individualistic lens, in terms of 'informational self-determination' or 'data autonomy', which often translate to 'everyone's right to control one's own personal data'. It is then (falsely) suggested that the need for data protection is satisfied if, for instance, consent is given with respect to the processing of one's own data, or if data are only collected anonymously. This individualistic and liberal framing of data protection in many public discourses neglects the data externalities – that is, the impact of our data on *other people* – that can result from anonymised and consensually collected data once available at scale. Indeed, the individualist framing of privacy, which makes the data subject responsible for the decision about the processing of their data, privatises the collective ethical problem and loads it onto the individual, just as individual-oriented ethics privatises collective effects in the form of individual moral deliberations and decisions. But the prime problem in an era of big data and predictive analytics is not individual privacy. The problem is that the owners of data and predictive models can *potentially* breach *anyone's* privacy as long as there is a significant portion of users who care solely or primarily about their *own* privacy.

Data protection measures that are restricted to concerns about individual control over information disclosure are thus toothless in the face of how predictive analytics is enabled by the data-sharing habits of a sufficiently large group of people. This lack of bite also applies to the General Data Protection Regulation (GDPR) of the European Union, for at least two reasons (Mühlhoff and Ruschemeier, 2022, 2024a). First, the GDPR does not apply to anonymised data, but anonymous data are sufficient for the training of predictive models. Referring to Figure 5.1 again, only the grey and white data pairs need to be present in the training data; a specific personal reference is not relevant and can be removed. Hence, anonymising the training data is in many cases a simple tool for circumventing the scope of the GDPR, which makes the training of predictive models a kind of data processing that is unregulated by current data protection laws. Moreover, in most cases, the trained model *itself* will also be counted as anonymous data. A predictive model (viewed as a matrix of weights and parameters representing the trained state of, for instance, a simulated neural network) is itself a form of highly aggregated data. If the training uses state-of-the-art anonymisation techniques such as differential privacy in machine learning,[8] the model data must be considered as anonymous data,[9] which can be freely processed (copied, distributed, sold or reused for other purposes), as these data do not fall within the scope of the GDPR. Hence, in general, the GDPR regulates neither the training nor the further processing of predictive models (see also Mühlhoff and Ruschemeier, 2024a).[10]

Second, the GDPR makes it too easy to lawfully collect training data from users. One of the most important legal bases according to the GDPR, 'consent', is a weak point because consent is relatively easy to obtain from users, even though in most cases it is neither voluntary nor informed in digital contexts. When a company seeks to collect usage data for predictive analytics, it is realistic that 'statistical evaluation of anonymised usage data' might be listed as a processing purpose. Many users would routinely consent to this, in particular given the promise of anonymisation. Moreover, it has been much discussed that dialogue boxes asking for consent tend not to inform users properly, but often trick or coerce them into giving their consent by means of design tricks, nudges, lengthy small print and because they are shown at inappropriate moments (see also Baruh and Popescu, 2017; Mühlhoff, 2018b). The principle of consent also does not fully encompass the collective enabling structure of predictive models. It hides the fact that by giving consent, the data subject is making a decision for many other individuals and, ultimately, society at large. Indeed, the current practices around consent that have emerged as a result of recent legislation, largely based on pop-up dialogues and cookie banners, contribute to the hegemonic misconception of what is at stake in data protection. Each new consent dialogue affirms the (false) liberalist imaginary that privacy is all about individual choices with respect to the sharing of (anonymised) personal data. Therefore, the current practice of consent is not only impotent, but potentially even *harmful* with regard to the data protection challenge of predictive analytics. It gives the comforting and erroneous impression that individual control is all we need, and thus it distracts public awareness from the collective dangers of data accumulation and predictive analytics (Sloan and Warner, 2014; Kröger et al, 2021).

Our current regulatory paradigm that governs the field of data protection has as its central area of concern the individual, their right to privacy or to informational self-determination, and thus their ability to control who can process their personal data. This is because privacy legislation is first and foremost constructed as the protection of fundamental rights against potential infringements arising from the processing of certain types of data. Fundamental rights are essentially subjective rights; they pertain to the individual. While the *effects* of predictive privacy violations still constitute potential infringements of subjective rights, these infringements cannot, in our age of big data and predictive modelling, be prevented by giving each of us control over who can process our personal (and anonymised) data. Our individual data practices have supra-individual consequences. To adapt the famous 1960s slogan 'The personal is political', we can now say 'The private is political', in the sense that your putatively *private* decision, relating to the kinds of data you share with platform companies, contributes to *collective* effects.

At the same time, it is no trivial task to update data protection legislation in order to tackle the collective effects of individual data disclosure.[11] I am asserting here that the notion of *predictive privacy* represents a promising ethical pathway in that direction. However, we should acknowledge that predictive privacy as defined in definition 1 is again a rather individualistic value: the concept is constructed in such a way that it allows us to identify a harm only to one specific individual. This definition is really about the predictive violation of *someone's* privacy. But, as discussed in this present subsection, two important insights must be reflected in any efficient data protection approach to predictive analytics. First, that harm is enabled by the serial behaviour of many – in other words, by all of our collective data-sharing habits. And second, the harm caused by predictive privacy violations that could result from a predictive model that was trained from our collective data could affect *anybody*, not only one particular individual. Accounting for these two extra features that make things so complicated, we need to understand predictive privacy more broadly and more collectively; we need to posit predictive privacy not only as an individual value, but as a societal level of protection against predictions (Mühlhoff, 2023b). Precisely because it is the collective nature of our shared behaviour that determines the level of predictive privacy that applies to all of us, we should construct predictive privacy 'as a protected good in the name of the common good' (p 7). Predictive privacy as a protected good then becomes the shared, common interest to protect all of us from the *potential* violation of our predictive privacy that looms as soon as a certain company develops or owns a predictive model. To reflect this collectivist perspective, a second version of the concept of predictive privacy might be given as follows:

Definition 2 (predictive privacy, positive version): Predictive privacy as a protected good designates a (legal and ethical) level of protection of the community against the predictive capacities of large data processors; that is, a demand for protection against a specific, contemporary, technological manifestation of informational power asymmetry (Mühlhoff, 2023b: 7).

Note that definition 1 provides a negative and individualistic definition of predictive privacy in its assertion that someone's *individual* predictive privacy is violated through predictions. Definition 2, in contrast, articulates predictive privacy as a *collective* value or interest. By 'collective' I envision, similarly to how Priscilla Regan has argued for privacy in general to be a collective value, that predictive privacy is not only a *common* value in that it is shared among individual people; it is also a *public* value in that it is vital to the democratic political system, and a *collective* value 'in that technology and market forces are making it hard for any one person to have privacy without all persons having a similar minimum level of privacy' (Regan, 2002: 399).

In the case of predictive privacy, this collective aspect of its value connotes that once a predictive model is created, it can be indiscriminately and automatically applied to many people – and potentially to any of us – at any time. The issue is the *potential* harms that may arise from applying a predictive model to *anybody*, as well as the scale and extent of those harms. This scale easily expands, as a single predictive model can be copied to many domains and applied to many people around the world. Even if these harms do not affect all of us at the same time (or in a similar way), preventing them is a collective interest because they could affect each and every one of us, and indeed *will* affect enough of us for their effects to become of structural importance.

This situation echoes the debate around prohibiting gun ownership. The risk of a weapon killing someone is too great to sanction only the use of the weapon; instead, and to safeguard our security, the production and possession of weapons must be regulated. This resonates with the stark words of the mathematician Cathy O'Neil, who describes a predictive model as a 'weapon of math destruction', which is a 'mathematical weapon of mass destruction' in the hands of a private company (O'Neil, 2016). This hyperbolic expression conveys the two critical points about predictive models: 'mass destruction' alludes to the fact that even a single predictive model can affect large swathes of society; the 'weapon' aspect alludes to data models' similarity to weapons in that they are both kinds of artefacts whose production and circulation – and not just their use – should be regulated, restricted and controlled for the good of all of us.

Regulating prediction power: why we need a preventive approach

In slightly less metaphor-friendly terms, an ethical debate and effective regulation of data-based AI technology must be concerned with the *power* that accumulates in the hands of the operators of this technology. The problem is already the *potential* violation of privacy or the *potential* harms of pre-emptive unequal treatment that can, and will, result from the use of such models. This makes ethical and political intervention an issue of prevention and precaution, implementing a 'protection-in-advance' approach that safeguards society before harm is done to specific individuals. As we are dealing here with outcomes that can potentially affect anyone, we need to address this problem as an instance of structural power. The concept of prediction power (see Chapter 5) describes the structural power that arises from the possession of predictive models. This relates to the ability of operators to apply these models automatically and at scale to make predictions about arbitrary individuals in a way that leads to differential treatment. This predictive capacity is coupled with the economic and political power of the operators – typically, large platform or tech companies. This combination allows predictions to be turned into material differences in the treatment of individuals which, in terms of potentiality, can affect anybody and society at large. Prediction power is thus the most recent manifestation of an informational power asymmetry between a few economic players on the one hand and individuals and society on the other.

If unregulated, prediction power easily multiplies and will likely lead to a further widening of social and economic disparities, discrimination and social selection around the world. To make clear at which point in the creation of predictive models prediction power arises, where the current regulatory gap is located and why this gap even enables prediction power to multiply, we now look at the typical data-processing chain that characterises the life cycle of a predictive model. This life cycle features (at least) three stages, each relating to different data subjects (see Table 6.2). To make the presentation of the stages more accessible, let's also use a hypothetical example. Imagine a psychiatry research group that plans to build a predictive model to predict incidences of depression by using speech data – that is, to use audio recordings of a patient's voice to predict the likelihood of that person suffering from a condition like depression.[12] The original idea is to improve medical diagnosis by such an AI system. We can look at three key stages of the life cycle of such a predictive modelling project[13]:

Stage 1 consists of collecting training data and training the predictive model. In our example, the clinical research group might ask existing patients to volunteer to participate in the study. Participants then give their consent for the recording of their voice as audio data and for the use of their (anonymised) medical treatment records. In general, as argued in the last subsection, training data are often easy to obtain legally – for instance, if consent is chosen as a legal basis. Obtaining this consent is often straightforward, especially when anonymisation can be promised to the data subjects. Hence, the GDPR does not effectively prevent this processing step, which is where predictive models are created from accumulated data. This is also the very step where prediction power is sourced, inasmuch as prediction power comes hand in hand with the possession of predictive models. The product of step 1 is a trained model that allows a prediction of depression in *any* individual based on recorded audio.

Moreover, for the subsequent argument it is relevant that the trained model can itself be considered as a specific kind of dataset, given by the

Table 6.2: The three processing steps in the schematic life cycle of a predictive AI model

	Step 1: Training the model	Step 2: Circulating the model	Step 3: Using the model
Input data	Training data, potentially anonymised	Trained model (to account for the more difficult case, I assume the model data to be anonymous)	Auxiliary data about the target person
Output data	Trained model as represented by the model data; in many cases, anonymous data	Copy or extended version of the trained model	Predicted information about the target person
Data subjects concerned	Individuals in the training data	None	Target person
Potential harms and effects	Aggregation of prediction power	Multiplication of prediction power, unaccounted transfer of models to new contexts, potential application of the model to anybody	Predictive breach of target person's privacy; discrimination, manipulation
Regulation that applies to this step	GDPR if training data contains personal data; none if anonymous data	None	GDPR

matrix of weights and other internal parameters of the trained model. In the general case, I am assuming that this data is anonymous, because this is the ideal scenario from the perspective of the operator, who would benefit from the trained model not falling within the scope of the GDPR if it constitutes anonymous data. In theory, and often also in practice, building an anonymous model is possible even if the training data are not anonymous. For this purpose, an array of recent anonymisation techniques, such as differential privacy for machine learning and federated learning, are available (see Chapter 6, Note 8).

Stage 2 then involves storing the trained model and processing it further – for example, by copying, selling, publishing or incorporating it as a subcomponent in an even larger AI model. (Using the model to calculate predictions is not part of this processing step.) Crucially, if the trained model is comprised of anonymous data, there are no existing legal restrictions on its distribution; it could be sold to other companies or released for free or as an open-source model. In our example of a model that predicts depression based on speech data, even if the original motivation of the project is to improve psychiatric diagnosis, the model could, for example, be shared with a company that develops AI systems to support hiring processes. Perhaps this company is interested in reusing the model as a subcomponent of its larger AI system that automatically analyses video-recorded job interviews. Such an unforeseen secondary use of the model may very likely lead to discrimination, exploitation and unfair treatment of already vulnerable groups. This constitutes not only a societal harm, but is also a clear breach of trust towards the donors of the training data in stage 1 who are members of this vulnerable group, hoping to contribute to improved medical treatment for their condition (Mühlhoff, 2024; Mühlhoff and Ruschemeier, 2024c). There are currently no effective restrictions on circulating a trained model to other actors and companies if its model data consist of anonymised data, which is, as stated before, a realistic technical assumption. As we see, stage 2 is the point of unchecked multiplication of predictive capacities due to a complete lack of effective regulation.

Stage 3, finally, constitutes applying the predictive model to a specific individual. In this stage, there emerges a new data subject who was potentially not involved until now: the target individual, who does not need to be one of the original training data subjects. Personal information about this target individual is predicted based on auxiliary data that are used as input data to the trained model. This is why the data processing in stage 3 falls within the scope of the GDPR such that all data protection provisions apply. However, there are many situations in which people are effectively coerced to consent to the use of predictive models when it comes to their case. For instance, when someone applies for a job, there is hardly a viable option not to consent to the automated processing and analysis of the application if this is demanded by the employer, as refusal would simply result in one's application not being considered. (We are not talking about high-profile jobs or management positions here, but about typical job interviews in what used to be termed the 'blue-collar sector', with a large turnover rate and with many more applicants than available positions.) In stage 3, where an individual, equipped with the defence rights enshrined in the GDPR, typically faces a large data-processing company, regulation comes too late as the power asymmetry is already manifest. The same imbalance of power applies in many cases of credit and insurance risk assessment.

Efficient regulation and control of the phenomenon of prediction power must therefore not solely rely on equipping the *target individual* with defence rights (which is the approach of the GDPR). After all, the set-up in stage 3 is an unequal fight because the affected individual is facing a company that is equipped with highly capable data analytics technology. This constellation is generally marked by a huge economic power imbalance, which is enhanced in this specific case by the accumulation of prediction power. Accumulation of prediction power on the part of the model's operator means, for instance, that the operator can afford to disregard applicants who *do not consent* to the use of predictive analytics in the assessment of their applications. In other applications such as differential pricing based on predictive analytics, prediction power could mean, for example, that the operator can afford to provide discount offers to anyone who consents to the use of the predictive model on their data. The market position and the competitive market strategy of the operator is profoundly shaped and stabilised by prediction power.

This is why stages 1 and 2 in the development of predictive models signpost a glaring regulatory gap. These are the stages at which prediction power emerges and multiplies. Hence, already the training, and not only the use, of predictive models needs to be regulated. If state regulation aims to mitigate power imbalances, this means regulating the production and circulation of predictive models in the case of predictive analytics. Definition 2 of predictive privacy as a common good is geared towards regulation of predictive capacities per se, *before* they are actually applied to anyone in step 3.

I have argued extensively in collaboration with the legal scholar Hannah Ruschemeier (Mühlhoff and Ruschemeier, 2022, 2024a, 2024b, 2024c) that in order to turn the ethical value of predictive privacy into law we need *preventive* regulation in the sense of the precautionary principle.[14] This preventive regulation must control and limit the emergence and circulation of predictive models. The dominant industry trend to make trained models freely available as 'open source',[15] which in the EU's AI Act even leads to exemptions from some provisions, is clearly a step in the wrong direction (Mühlhoff and Ruschemeier, 2024c). One of our regulatory proposals, purpose limitation for models, demands that the training of predictive models should be limited to purposes that are compatible with the purposes for which the training data were collected from the training data subjects (before they were potentially anonymised), and that the use and reuse of both trained models and training data (even if anonymised) should be limited to these purposes (Mühlhoff and Ruschemeier, 2024b, 2024c).

We argue that a regulation like this is vital not only to prevent large-scale harm due to the misuse of predictive capacities for discriminatory purposes, but also to encourage trust in, for instance, medical applications of AI. Again referring to the example of disease prediction from speech data, such a project would require thousands of patients volunteering to donate their data (medical records and recorded audio data) to be used as training data. Many patients are generally motivated to donate their anonymous data to medical research, hoping to contribute to a future improvement of diagnostics and therapy. Donating data to such research in fact serves the common good, as potentially anybody might benefit from resulting medical innovations. But in the current state of (missing) regulation, there is as yet no guarantee for the training data donors (or for society at large) that the AI models trained from this voluntarily given data might not get reused for unintended and potentially harmful purposes like insurance risk scoring or automated hiring decisions that in the end discriminate against people with current or potential health conditions. To ensure trust in beneficial

AI research and applications such as in medicine, there should therefore be a legally enshrined mechanism to ensure that training data and models can only be used for the intended purpose that needs to be explicitly stated before the model is trained. With such a purpose limitation for models, we intend to prevent the risk of unaccounted reuse for other purposes, which is a risk that could manifest even years later.

7

The Culture of Prediction:
Ethics and Epistemology

Predictive analytics is one of the largest domains of machine learning applications today. Leveraging collections of data on many individuals or cases, predictive analytics allows for the derivation of unknown and potentially sensitive information about new cases. This technique is routinely and in many areas used to make decisions about individuals (such as in algorithmic hiring), to calculate risk scores (such as in the insurance industry) or personal recommendations (as in targeted advertising), or even to make predictive sentencing decisions in the criminal justice system. Due to the widespread use of predictive AI technologies across sectors, including research, policy and economics, we may speak of a new epistemic culture, or culture of prediction. As argued in the last chapter, predictive analytics leads to serious ethical questions in relation to the dignity and autonomy of affected individuals. To elucidate these ethical questions in more detail, we need to dig deeper into the philosophy and epistemology of statistics to better understand the characteristics that mark the culture of prediction.

This will be the mission of this chapter. First, I will point out that predictive modelling for decision purposes is marked by an epistemological and ethical problem that I call the 'prediction gap'. I will then argue that this problem is more broadly a manifestation of a profound shift in the underlying epistemic framework of statistical knowledge production that has happened over the past decades. This shift involves a transition from classical 'frequentist' statistics – with its 'objective' notion of probability – towards what is often termed 'Bayesian' statistics, involving a so-called subjective notion of probability. Finally, I will briefly debate what is ethically and politically at stake with this transformation in the ways we produce statistical knowledge.

Inference vs prediction: the prediction gap

A predictive model rarely makes an unambiguous decision; rather, the output is typically a list of possible values of the target variable, given probability weights.[1] For example, a predictive model for criminal recidivism might calculate a probability of 0.65 for 'recidivism' and 0.35 for 'no recidivism' in a specific case. If a decision about a defendant is to be made on the basis of such a prognosis, however, a *specific* value must be selected from the range of all *possible* values of the target variable – the person cannot be released and incarcerated at the same time.

An obvious solution is to choose the value with the highest probability weight. In the example, this means treating the person with a recidivism weight of 0.65 *as if we know for sure* that they will reoffend. At this moment in the chain of reasoning we cross what I call the 'prediction gap': as soon as real action follows from a prediction, the ambiguous outcome of a statistical assessment is disambiguated by choosing *one* of the possible options. The person is being pinned down to a behaviour that seems *most likely* compared to many other known cases – and the person is thus denied the possibility of being an 'outlier' from that statistical analysis. Ethically speaking, this treatment denies the person autonomy, freedom of action and a principled diversity of human manifestations. After all, the relevant behaviour is completely down to that person's own volition.

The prediction gap thus marks out a logical, epistemological and ethical problem. This problem occurs when what is actually a distribution of probability weights over a range of possible outcomes (an answer to 'what could happen?') is simplified to a point prediction: an unambiguous answer to 'what will happen?'.

In a classical understanding of statistics,[2] statistical analyses of large sets of sample data lead to what is called a 'statistical inference'. According to a standard dictionary of statistics, statistical inference is: '[t]he process of drawing conclusions about a population on the basis of measurements or observations made on a sample of units from the population' (Everitt and Skrondal, 2010: 2017). That is, a statistical inference represents laws or regularities, expressed in probabilistic terms, that pertain to the *full population* of individuals or cases from which the sample data were obtained (for example, smoking prevalence correlates with the rate of developing cancer in a specific population). Such an inference constitutes aggregate knowledge about a large set of cases or individuals and not yet a prediction about a specific case.[3]

Hence, in the orthodox understanding, a statistical model trained from sample data constitutes a statistical inference. That is, it describes statistical knowledge that *generalises* from the sample covered by the training data to some kind of population-level knowledge. In predictive analytics, then, that statistical model is abused for case-based predictions and decisions. The statistical inference is taken not as knowledge about what happens at a population level (for example, on average, smoking causes cancer), but is applied to a single case to obtain knowledge about a particular individual (for example, Tina smokes, hence we predict that she has a high chance of developing cancer and as a consequence we treat her differently from lower-risk people).

Frequentist vs Bayesian probability

In the framework of classical statistics, from Tina's smoking we cannot derive the *fact* that she will develop cancer. This is because large areas of 20th-century statistics rely on a 'frequentist' conception of probability that interprets probability as the relative frequency of an event in a hypothetical infinite number of repetitions of the same experiment or scenario. When we say that a coin shows heads with a probability of 0.5, in the frequentist understanding, this means that in a series of infinite (and independent) coin tosses, the counts of heads and tails will even out. A similar interpretation holds for 'smoking causes cancer' in the frequentist sense: having an arbitrarily large number of people who either do or don't smoke, the number of those who will develop cancer is higher among smokers compared to non-smokers, if all other relevant parameters are randomly distributed. Those statements pertain to the limit case of the number of cases going ad infinitum, but they do not afford, in the frequentist interpretation of probability, any information about an individual case. In the classical paradigm of statistics we only see the big picture of population-level correlations; we are not able to focus on any particular individual or case.

Crossing the prediction gap marks the moment when a statistical inference, which is knowledge about a population, is turned into a point prediction – that is, into predictive knowledge about a specific individual or case. This step, which is at the heart of predictive analytics and contemporary machine learning applications of all kinds, requires an altered epistemological understanding of probability (see also Joque, 2022). It is not useful to define probability as the relative frequency of an event when we aim to make a judgement about an individual or case in a specific situation in a way that triggers real and material consequences, because such a situation means dealing with profoundly non-repeatable cases. Predicting whether Erin suffers from substance abuse and basing the decision about whether or not to invite her to a job interview on that prediction is not a scenario that allows for infinite independent repetitions in any meaningful sense. The same holds true when a local weather forecast says it is 70 per cent likely that it will rain tomorrow in my town. We understand this figure as a statement about a specific day (tomorrow, for instance), in order to decide on the best form of transportation to use to get from A to B during that time period. This statement is only useful as a statement about this one instance – tomorrow – and not as a statement about the relative frequency of rain in a fictitious infinite series of independent repetitions of the day that is tomorrow.

This is where Bayesian statistics comes in, which is an older school of statistics attributed to Thomas Bayes (1701–61), although it was extensively developed only later, by Pierre-Simon Laplace (1749–1827). Bayesian statistics operates with a conception of probability that applies to the single case or event rather than repeated trials, in the sense that it quantifies a *degree of belief* in some outcome of a process against the backdrop of incomplete information. In a typical scenario involving this notion of probability, a judgement or decision is to be made about a singular case, in a situation where the observer, who is also the decider, does not have the means to fully determine the outcome of the process in question from their subjective viewpoint. This means that the decision has to be made based on 'rational belief' about the outcome. Quantifying this belief as a Bayesian probability value can be interpreted in the context of gambling strategy as corresponding to the amount of money one would rationally bet on the correctness of the belief. If it is 30 per cent likely that it will rain tomorrow, what extra costs of going by car instead of by bike would be rational to pay, provided it is of immense importance to me to not get drenched on the way? Or, to reveal the cynicism behind this kind of reasoning, if a predictive model says Erin is 20 per cent likely to suffer from substance abuse, which might severely impair her work performance, how much would I bet on her becoming a 'productive' member of staff and, correspondingly, which resources would I 'rationally' be willing to invest in her hiring, training and onboarding?

The difference between frequentist and Bayesian probability is a difference between an objective vs subjective knowledge claim. In the frequentist paradigm, probability is attributed as a property to the objects that are involved in the experiment (for example, a coin being tossed, a human body, an experimental set-up); dependence on the person who wonders about that very probability (that is, the observer, the decider) is eliminated through the assumption of infinite independent repetitions that are merely *passively* viewed by the observer. In the Bayesian conception, probability lies with the observer, as the need for probabilistic reasoning arises when information is incomplete but a rational 'best-bet' decision needs to be made by the decider with regard to the process in question.

This shows that the difference between frequentism and Bayesianism is not merely a change in perspective, but that different problems can be put into focus with both notions of probability. The frequentist conception of probability is inherently tied to the history of experimental sciences such as physics, genetics and medicine, whose knowledge production relies on the general epistemological assumption that objective empirical insight can be directed to a circumscribed (and small) compartment of space and time – the scene of the experiment – that is untouched by the external observer (experimenter) and in principle eligible for infinite repetitions or replications.[4] The Bayesian conception of probability, on the other hand, does not rely on such a characteristic epistemological subject–object split. It is applicable to situations where interest is not on objective knowledge about a case, but on the pragmatics of reliably and 'rationally' judging a case based on a limited number of similar observations. Despite these different strengths and weaknesses, frequentist statistics has a quasi-hegemonic status in 20th-century science, with undergraduate textbooks and large parts of empirical methodology in psychology, biology and medicine almost exclusively relying on this paradigm. However, during the past 20 years, a tacit transition in our culture of empirical knowledge production towards a subjective and pragmatic understanding of probability is at the heart of the machine learning revolution (Joque, 2022). When predictive modelling is used to decide cases, rank options or score risks, the concept of probability that is involved no longer refers to relative frequencies in large numbers of trials. Rather, these applications try to answer the question of which outcome of a process is the rational 'best bet' given the available information, so that the case can be treated in an economically efficient way.

The new culture of algorithmic modelling

Since the turn of the millennium, the difference between frequentist and Bayesian conceptions of probability has been marking a deep research–political schism, inasmuch as the former presents the dominant paradigm of institutionalised statistics in the 20th century and the latter informs the comparatively new approaches that today are known as machine learning. The US American statistician Leo Breiman points out this schism in a combative 2001 paper titled 'Statistical modeling: the two cultures'. In this paper he accuses the research field of statistics of being stuck in a rather old-fashioned approach of 'data modeling', which forms an orthodoxy shared by 98 per cent of scholars in the field. Scholars and research domains stuck in data modelling, according to Breiman, fail to see the potentials of the emerging paradigm of 'algorithmic modeling' (Breiman, 2001: 199). Since the 1980s, the latter of these two opposing 'cultures' has constituted a 'new research community' outside traditional statistics departments that has been busy developing machine learning techniques such as 'decision trees', 'support vector machines' and 'neural nets' (Breiman, 2001: 205).

Beyond its observations on the politics of statistical research, Breiman's paper presents an interesting description of the different epistemological frameworks of the two modelling cultures, which complements the difference between frequentist and Bayesian understandings of probability. Breiman describes the general starting point of statistical modelling, shared by both cultures, as follows: 'Statistics starts with data. Think of the data as being generated by a black box in which a vector of input variables x (independent variables) go in one side, and on the other side the response variables y come out' (Breiman, 2001: 199).

The notion of 'black boxes' that represent processes in 'nature' that are to be statistically modelled is central to Breiman's presentation, as he points out that the two 'cultures' of statistical modelling differ in the way they approach these black boxes. While data modelling tries to elucidate and eventually 'fill' the inside of the box (p 199), algorithmic modelling only 'emulates' the box's behaviour (p 204).

The goal of the data-modelling approach is to use the available data pairs of inputs x and responses y to find a formula (more precisely, a 'model') that can be substituted for the inside of the black box that represents the probabilistic process in question. This could be, for instance, a linear regression model, a logistic regression model or a Cox model (p 199). This procedure of finding a specific formula that fills the inside of the box as it allows for the calculation of y from x, however, depends on assumptions regarding the general structure of that formula. For instance, by selecting linear regression, a *linear* relationship between x and y is assumed; in the case of logistic regression, linear combinations of the independent variables x are used to calculate probability statements about y by filtering it through the logistic function. In Breiman's words:

> The analysis in this culture starts with assuming a stochastic data model for the inside of the black box. For example, a common data model is that data are generated by independent draws from response variables = f(predictor variables, random noise, parameters). (Breiman, 2001: 199)

So, the data-modelling process starts from assuming a specific mathematical structure of the formula f that represents the inside of the box. Its first argument, the 'predictor variables', are the independent variables x that serve as input to the process. The second argument, 'random noise', is some random parameter that accounts, for instance, for varying responses y to the same x if the same x is run multiple times through the process. The third argument, the model 'parameters', are constants by which the general structure of the formula f is *fitted* to the specific data pairs x, y on which the data modelling is conducted. Hence, the data-modelling procedure consists of estimating the model 'parameters' from the available data by means of statistical 'fitting' methods. A 'goodness-of-fit' test is usually employed to decide whether a fit is valid or not (often by a 5 per cent level of significance). All empirical scientists know this by heart as 98 per cent of them are dealing with this methodology, according to Breiman.

For example, let's imagine the 'black box' that we want to model is the dependence of body weight on body height in people from a certain cohort. The basis for our statistical analysis is several data pairs of height (x) and weight (y) values of existing individuals. Approaching this with a linear regression model means that we assume a relationship of the form $y = a + bx + e(x)$, where a is the offset and b the proportionality constant of the assumed linear relation between x and y, and where $e(x)$ is a random noise component. Fitting this model on the data would then be a matter of determining the specific values of a and b and ascertaining whether they pass a 5 per cent confidence test. Notice that assuming a linear relationship between height and weight is brought to this analysis as an a priori hypothesis. Often, such hypotheses are anchored in theoretical reasoning or previous empirical findings, for instance, such that we might expect that body weight could be in some way proportional to height.

Breiman presents the other culture, algorithmic modelling, as taking a different approach towards the back box. Its practitioners are not striving to find a formula (model) that (approximately) represents and explains the inner mechanism of that box, but to build another – potentially complex and hard-to-interpret – black box that only 'emulates nature's box' (p 204). Key to this approach, according to Breiman, is the transition away from significance testing with respect to the model parameters and towards the 'predictive accuracy' of the emulation: 'put a case x down nature's box getting an output y. Similarly, put the same case x down the model box getting an output y'. The closeness of y and y' is a measure of how good the emulation is' (Breiman, 2001: 204).

The 'model box' being constructed in this approach is now an 'algorithm' rather than a 'data model'. What exactly is the difference? After all, the data model can also be used to calculate predicted y' from given x.[5] The difference is that algorithmic models do not, as data models do, commit to a rather narrow a priori assumption regarding the mathematical relationship between input and output variables (for example, linear combinations). Algorithmic modelling, according to Breiman's juxtaposition, avoids any preassumptions about the mathematical or causal relationship between input and output variables, and instead looks for 'the models that best emulate nature in terms of predictive accuracy' (p 209). Algorithms can, after all, be much more complex (for example, by featuring various interdependent or iterative steps) than what can be expressed in a simple $y = f(x, noise, parameters)$ formula.

This difference has important consequences for interpretability. A data model, inasmuch as it operates with assumed mathematical structures (formulas) governing the relationship between input and output, brings with it a notion of interpretability. Fitting the relationship between body weight and height with a linear model like $weight = a + b \times height + noise$ must be grounded on the reasoning that, as bodies get taller, their width does not change to an equal degree, because otherwise the relationship would be a quadratic one. The presupposition of a certain data model f amounts to the presupposing of a certain causal or theoretical model governing the coupling of the independent (input) variables with the dependent (output) variables. This approach starts from an abstract assumption and turns to the data to fit the specific model parameters that represent the numerical strength of the diverse (linear) causal couplings that are part of that model.

The algorithmic routines that constitute an algorithmic model cannot generally be represented in an interpretable mathematical formula that maps inputs to outputs. The idea that we should start from an abstract assumption of what the coupling of x and y looks like is dropped in favour of an approach that turns to the external – you could say behaviourist – criterion of predictive accuracy. The result of this modelling procedure can be an algorithm of whatever kind as long as it produces, according to Breiman, 'useful', 'reliable' and 'accurate' information about the output that is to be expected from a given input (see also p 210). 'The goal [in algorithmic modelling is thus] not interpretability, but accurate information' (p 210).

The loss of interpretability is widely debated and often critiqued in relation to the use of machine learning models today (Matthias, 2004; Burrell, 2016; Mittelstadt et al, 2016; Pasquale, 2016; Samek et al, 2017). As we see in this brief historical excursion, this 'loss' refers to the fact that algorithmic modelling does not *start* from theoretically motivated, a priori assumptions about a general mathematical structure that represents the interior of the black box. Instead it adopts a kind of behaviourist approach that is uninterested in representing the 'interior' of the box and instead strives to build an algorithm that 'reliably' predicts the outcome of the process in question. Hence, the lack of interpretability is related to reversing the traditional order of 'theory first and data analysis second', with data analysis first and theoretical interpretation second (or even omitted) in the machine learning paradigm. The extension of this problem even to the social sciences has been famously criticised as 'the end of theory' – as the production of empirical knowledge that is no longer theory-guided in terms of statistically testing hypotheses (Anderson, 2008; Kitchin, 2014; see also Chapter 9).

Before we all bemoan the death of descriptive and interpretive modelling and the 'scientific method' (Anderson, 2008), we shouldn't forget that the frustration with data modelling is largely based on the limitations and simplistic assumptions of that approach in practice. Few available data models such as linear regression, logistic regression and the like make up the main part of the models that are routinely and often unreflectively applied in many empirical investigations. It is thus important to see that even the data-modelling culture only offered rather *limited* interpretations that were prestructured by the a priori assumptions about the model that was being used. As Breiman puts it:

> Usually, simple parametric models imposed on data generated by complex systems, for example, medical data, financial data, result in a loss of accuracy and information as compared to algorithmic models [...]. There is an old saying 'If all a man has is a hammer, then every problem looks like a nail.' The trouble for statisticians is that recently some of the problems have stopped looking like nails. (Breiman, 2001: 204)

Based on this paper from more than 20 years ago, I maintain that the success of machine learning has brought about a profound transformation of our empirical knowledge culture that stretches across many scientific, economic and political domains. This shift is motivated by ever newer and more complex problems, and a fresh wealth of data that enable statistical analyses where the decades-old repertoire of standard data models no longer suffices. Think of speech recognition, image classification, machine translation – it is hard to imagine resolving these problems with a data-modelling approach. With the shift from data modelling to algorithmic modelling, however, we also see how the epistemic interest shifts away from the descriptivist and objectivist pursuit of gaining universal theoretical insight. What matters more today is building machines that enable their situated and subjectively interested operators to reliably 'emulate' the process in question. The promise is for the operators and users of such machines, who are typically in situations of having (access to) limited information, to look ahead in time in order to manage risks and foresee opportunities. This epistemological interest is much closer to economics, game theory and gambling than to descriptive and universalist science.

Automating Sherlock Holmes and the ethics of statistical reasoning

With today's proliferation of data analytics and machine learning, we are now witnessing a fundamental transformation of our knowledge culture (see Table 7.1). But is this only a transformation of our *knowledge* culture? The knowledge production that makes use of predictive modelling is usually embedded in practical contexts in which action follows from that knowledge – and the knowledge is, in turn, entangled with action. This is evident in the many examples where predictive knowledge is used for automatic or semi-automatic decision-making that amounts to treating people differently – from the ads people see to criminal sentencing decisions.

Table 7.1: Comparison between inferential and predictive knowledge production

	Inferential knowledge production	Predictive knowledge production
Statistical paradigm	Frequentist statistics, objective probability, data modelling	Bayesian statistics, subjective probability, algorithmic modelling
Epistemological interest is in knowledge about ...	Total population	Individual/case
Empirical principle	Statistical generalisation, testing a mathematical hypothesis that represents the interior of the 'black box'	Best bet about individual/case, reliable emulation of the behaviour of the 'black box'
Availability of data	Data collection is expensive (as in, empirical research using questionnaires on a sample with)	Data are widely available (as in, social media usage data with $n = all$)
Mathematical problem	'How small can we make the sample chosen so that we can generalise to the total population with a given confidence?'	'Given data about a large part of the population, how can we gain knowledge about the most efficient treatment of a particular individual?'
Epistemic criterion	Objective truth (about hypothesised causal relations)	'Predictive accuracy' (Breiman) / 'predictive performance' (see Chapter 9)

Hence, the knowledge culture of predictive modelling becomes evident as a nexus of knowledge and *power* – that is, as a power-knowledge regime. The predictive power-knowledge regime leads to a transformation that is at the same time epistemological, ethical and political, and it is spreading across diverse realms, including those of science, economics, policy, education, welfare and administration.

Using a term by Katherine Hayles, we could say that predictive power-knowledge forms the core of the contemporary 'computational regime' of politics, science and society (see also Hayles, 2005). As Louise Amoore puts it, in this computational regime, 'what matters to the algorithm, and what the algorithm makes matter, is the capacity to generate an actionable output from a set of attributes' that are known about the particular case (Amoore, 2020: 4). Reliability of a prediction means reliability of the prediction as something that unambiguously translates into *action*. As we saw in the discussion of the prediction gap, this is all about controlling the future by disambiguation: the computational regime makes 'a political claim on the future', a future-grab that 'is effaced by algorithms that condense multiple potential futures to a single output' (p 4). It is thus the central, quasi-mythological promise of the computational regime of predictive modelling that computation relieves those responsible of the burden of dealing with ambiguity and uncertainty. Making a judgement in an essentially ambiguous context is being rationalised and mechanised by the Bayesian framework of algorithmic modelling, so that responsible individuals can emphasise a belief in the 'objectivity' and 'truth' of their approach. The responsible, empathic, judging human intellect is replaced by a culture of rational 'betting'. Questions such as 'Do I believe in Erin becoming a productive member of my workforce, despite her predicted 20 per cent likelihood of suffering from substance abuse?' are being transformed into a procedural kind of gambling: 'Let's have a computer calculate how much we should rationally bet on Erin as the best new hire'. Trust in human judgement and responsibility, as well as in personal commitment in social relationships, is under heavy pressure today. In line with these observations, the debate around 'automation bias' shows that there is a growing proclivity in our contemporary culture to trust automated and data-driven decisions more than human ones (Skitka et al, 1999).

As noted before, the prediction gap marks the point when the generally ambiguous output of a statistical model applied to a specific case is transformed into an unambiguous 'actionable' output. This is the moment of mechanised betting. While the model says Erin is 20 per cent likely to suffer from substance abuse and 80 per cent likely not to, we can only treat her on an either/or basis, hence we bet on 'too risky' and, by not inviting her to the job interview, treat her *as if we already know* that she would not be a reliable member of staff – although it is perfectly possible that she will not start abusing substances, not least as a result of her own volition (and perhaps even of getting the job). A fundamental ethical problem in crossing the prediction gap is related to this structure of making a bet *about* Erin instead of developing a committed relationship *with* her as a person. This reduction to a betting approach renders Erin as a mere factor of risk and contingency that needs to be managed, instead of a human being with whom we can build a relationship, whose motivations and outlook in life we might co-shape as her employer, and who, as a human being, must in principle be assumed to be free to decide on her actions, including to abstain from substance abuse.

Predictive modelling in the machine learning culture relies on (training) data that represent many other (known) cases and individuals. Hence, it is important to bear in mind that applying predictive models to a specific case, leading to a prediction about that case, involves implicit comparison of the case with all the data points in the training data. What is at work here is, in the end, pattern matching. Predictive modelling compares the individual to many other individuals who, given the same input data, seem statistically similar. Cathy O'Neil calls this the 'people like you' principle: we don't know *you* as an individual, but *people like you*, because, as the pattern matching of your data traces with the data traces of many others shows, you're generally more likely to develop substance abuse, hence it is our *best bet* that *you* will also slip into substance abuse.

Individuals are stripped of their autonomy and dignity when they are judged and managed according to such a 'people like you' scheme that, like Sherlock Holmes, suggests that 'we know already what you are like' (see also Basu, 2019; quoted after Mühlhoff, 2021). The US American philosopher Rima Basu makes a compelling case that this epistemic attitude of 'statistical reasoning' about people is 'morally objectionable' (Basu, 2019: 6). It constitutes 'a way of looking at another person not as a person, but as an object that is determined by causal law, as something whose behaviour is to be predicted' (Basu, 2019: 8). Based on this, Basu makes the important point that this kind of wrong treatment begins at the *epistemic* level of what we believe of one another. It is not only action that can be a moral wrong to another person, but people 'can also be wronged by *what is believed of them*' (Basu, 2019: 2, emphasis in original).

We have seen in this chapter that there are fundamental reasons why summoning people into the computational regime of predictive modelling is ethically objectionable. I call these reasons 'fundamental' as they do not pertain to faulty implementations of predictive technology, but pertain to the very idea that predictive judgements about individual people should be made in the first place. As I have argued, it is unfair, dehumanising and a violation of individual autonomy to treat people based on a statistical wager about their future behaviour that is derived from a cross-comparison with (the data of) many other people. We therefore urgently need an ethical debate that protects individuals against the pre-emption of their free choices and denial of their self-authorship of behaviour by the forces of predictive reasoning.

PART III

The Power of Control

8

AI Cybernetics

In this chapter I will link the debate around predictive modelling to the philosophical framework of Human-Aided AI in order to make the point that the capability to compare, classify and predict human behaviour is actually an *inherent* feature of Human-Aided AI systems. As the example of the constantly retrained and recalibrated ad-targeting algorithm shows (see Chapter 6), the most efficient predictive systems rely on real-world feedback. As I will discuss in this chapter, those systems are not only *sociotechnical* but indeed *cybernetic* apparatuses that sustain their operations through continuously measuring how users respond to the predicted content. Characteristic for cybernetics is the reliance on carefully designed feedback loops, which is evident in a wide range of mundane examples: in advertising and search results, click data are used as feedback data. In product recommendations, whether the user clicks on, shares or even buys a recommended product is used as feedback data. In the case of dating sites, matches (that is, recommended profiles of other users) are personalised predictions; feedback on these predictions is obtained by recording whether the user 'likes' the suggested profile in turn, whether the two start chatting, exchange phone numbers or even arrange a date. Services like YouTube, Netflix and Spotify, which offer personalised playlists or an infinity of 'next videos', create feedback loops by registering whether the user actually consumes the predicted content, or even bookmarks or shares it, or presses 'Next' instead.

In all these cases, feedback is implicitly established through the ability of digital media to capture our reactions to the content we see or hear. With the collecting of these data, a continuous relationship is established between predictive algorithms and the behaviour of actual users, designed so that the predictive model gets better with each additional point of usage data we leave behind. However, when considering these feedback mechanisms, it is crucial to analyse them as bidirectional. Not only are we users 'aiding' (and, in fact, enabling in the first place) these Human-Aided AI systems to make better predictions, but these AI systems are also acting on us. As much as those models rely on constantly observing human behaviours and reactions, they also affect and prestructure what we see and know, think or feel, how and where we move, with whom we communicate, which new music we discover or what gets our attention.

So, as we shall see in the course of this chapter, predictive systems, which are Human–Aided AI systems, not only passively learn from their human subcomponents by analysing their data, but simultaneously govern them in a process of *control through feedback*. The principle of feedback points to a specific modus operandi of *power* in the relationship between predictive AI systems and humans that controls individuals in an often subtle and indirect way – as we know, for instance, from the contemporary debate about 'nudging' (see Chapter 3 and Thaler and Sunstein, 2008; Yeung, 2017; Mühlhoff, 2018b). As I will point out in a brief historical excursion, this principle of governance through feedback loops owes a lot to the transdisciplinary science of cybernetics that emerged in the mid-20th century through seminal contributions by Norbert Wiener, Margaret Mead, Ross Ashby and others.[1] In fact, contemporary predictive AI can be traced back to cybernetics research related to military technology in the Second World War. The emergence of Human–Aided AI in the context of contemporary social media, then, appears as the latest historical stage in the development of cybernetics (see also Hayles, 2008).

A brief genealogy of predictive targeting in the history of cybernetics

For Wiener, 'feedback is a method of controlling a system by reinserting into it the results of its past performance' (Wiener, 1989: 61). In terms of what exactly a 'system' is, most authors in the field of cybernetics commit to a functional perspective that does not make essential distinctions between, for instance, physical, biological or social systems; Wiener famously speaks of cybernetics as a theory of 'control and communication in the animal and machine' (Wiener, 1948b). At the core of cybernetics is the principle of information feedback, by means of which cybernetics explains control processes that govern technical, biological and social systems, including communities, organisations and societies. Systems are understood as complex, self-organising entities that are not static but are constantly changing and adapting in response to their environment.

The ideas of cybernetics can enrich our analysis of Human–Aided AI systems and their specific form of power, because Human–Aided AI systems *are* cybernetic systems. To point this out, it is worth taking a brief look at the prominent role feedback loops play in some historical examples of cybernetics research. In its simplest form, the idea of cybernetics can be illustrated by the example of the heating thermostat: the cybernetic method of maintaining a certain target temperature in a room consists of constantly measuring the current temperature, comparing it with the target temperature, and turning off the heating if there is a positive deviation and turning it back on if there is a negative deviation (Wiener, 1948b: 96–7). This feedback loop is effected by the well-known mechanical or electromechanical devices called thermostats. This mechanism of a feedback controller (also known as a closed-loop controller) for maintaining a desired temperature is much simpler than blind 'open-loop' control, in which the heating power is set statically and without feedback, implying complex calculations of the required heating energy per time period based on the difference between the indoor and outdoor temperatures.

Cybernetics gained prominence in the 1940s as an engineering approach tested in military technology. During the Second World War, Norbert Wiener himself worked on a programme to improve the targeting of anti-aircraft guns and developed the idea of control through feedback in this domain. Hitting a distant moving target such as an aeroplane is complicated by the fact that the projectile requires a certain amount of time between firing an object and that object arriving at the target. The gun must be aimed not at where the target is at the moment of firing, but at where it is projected to be when the projectile arrives. A 'fire-control apparatus' was thus invented that was 'capable of tracking the curving course of a plane and predicting its future position' (Wiener, 1948a: 14). The difficulty of this task lies in the fact that the target's course may unexpectedly change, primarily due to evasive manoeuvres by the target aircraft. As Wiener recounts: 'We soon came to the conclusion that any solution of the problem must depend heavily on the feedback principle, as it operated not only in the apparatus but in the human operators of the gun and of the plane' (Wiener, 1948a: 14).

Building a fire-control apparatus therefore amounted to the 'design of prediction machines', as Wiener puts it (Wiener, 1989: 61). What this machine would need to predict is not so much the deterministic physical processes of how an aeroplane's trajectory is affected by wind and weather, but what evasive manoeuvres the human controllers are most likely to take. In his words:

If the plane were able to take a perfectly arbitrary evasive action, no amount of skill would permit us to fill in the as yet unknown motion of the plane between the time when the gun was fired and the time when the shell should arrive approximately at its goal. However, under many circumstances the aviator either does not, or cannot, take arbitrary evasive action. He is limited by the fact that if he makes a rapid turn, centrifugal force will render him unconscious; and by the other fact that the control mechanism of his plane and the course of instructions which he has received practically force on him certain regular habits of control which show themselves even in his evasive action. These regularities are not absolute but are rather statistical preferences which appear most of the time. (Wiener, 1989: 61)

Predicting human actions and reactions is thus at the core of this feedback-based, technological apparatus by which anti-aircraft fire predicts the target's future trajectory. The human behaviour of the aircraft pilot, shaped as it is by training, habit, psychology and physical constraints, is the main reason why a 'prediction machine' – as opposed to deterministic calculations of the plane's future position – is needed. Control through feedback loops opens up new possibilities for the mechanical, electronic or – later in the history of cybernetics – digital capturing of human behaviour.

Wiener's 'wartime project' is an origin story of the predictive modelling of human behaviour *for targeting purposes*, rather similar to what is practised today in networked media with respect to user behaviour. What is special about cybernetics is that it does not draw an essential distinction between the physical and the social, and thus can model complex sociotechnical assemblages according to the same principles. As we see, in the universal paradigm of cybernetics, it is not a big leap from *targeting* enemy aircraft to *targeting* users with advertisements, product offers, dating site matches or the next video. Both the military and the civil applications are manifestations of the same idea of cybernetic 'prediction machines' that learn from feedback. Indeed, according to some critical scholars, cybernetics is the central principle of the late form of capitalism based on communication, information circulation and control called 'cybernetic capitalism' (see also Tiqqun, 2020; Ström, 2022).

This genealogical excursus exemplifies how contemporary predictive analytics are historically rooted in the military technology of predictive targeting, which is a prime case upon which the broad scientific paradigm of cybernetics was developed. Conversely, this genealogy also suggests that the emergence of Human-Aided AI, particularly in the form of predictive AI systems, can be seen as the next stage in the proliferation and development of cybernetic technologies of control and governance. If the first thesis of Human-Aided AI (see Chapter 2) is that the feedback loops established through networked media, as well as human users themselves, are to be seen as constitutive components of AI apparatuses, this idea is already hinted at in Wiener's insights into the nervous system:

The feedback principle introduces an important new idea in nerve physiology. The central nervous system no longer appears to be a self-contained organ receiving signals from the senses and discharging into the muscles. On the contrary, some of its most characteristic activities are explainable only as circular processes, traveling from the nervous system into the muscles and re-entering the nervous system through the sense organs. This finding seems to mark a step forward in the study of the nervous system as an integrated whole. (Wiener, 1948a: 14)

This understanding of the nervous system parallels that of AI systems in Human-Aided AI. Just as the functioning of the nervous system can only be explained in terms of closed loops between central components and the sensorimotor apparatus to which it is bidirectionally connected, so too must data-driven AI systems be explained in terms of circular processes – that is, feedback loops that bidirectionally connect data-centre algorithms with processes in the world.

Cybernetics of social structures, or the performativity of predictive AI

But let's not stop at this historical analogy between contemporary predictive AI systems and cybernetic systems of the 20th century. It's worth looking a little closer at the idea of feedback processes as an essential component of Human-Aided AI systems, because it raises the question of who is actually in a mutual feedback loop with what? Is it the data-based representation of a user with the movements of that individual user (individual-level feedback loops)? Or is it the 'patterns' that the machine learning model recognises from observing and comparing *multiple* individuals, coupled with the patterns and structures of society (aggregate- or system-level feedback loops)?

According to the second thesis of Human-Aided AI, problems in AI 'correspond' to problems in user experience design (see Chapters 2 and 3). This shows that interface design is precisely the discipline that is occupied with the designing of feedback loops in AI. It is the immersion across graphical user interfaces that makes for hybrid circular couplings in which computational processes that run on silicon-based processors rely deeply on the personal and social practices of human users. This, in fact, leads us to a more refined way of stating the Human-Aided AI hypothesis: most commercial AI systems today are not merely sociotechnical but *cybernetic* apparatuses, enabled through contemporary media culture. These apparatuses rely on constant feedback between computational processes and human cognitive and social capacities.

Cybernetics, as a transdisciplinary endeavour, has always sought to apply the principle of control through feedback in the social sciences as well – for instance, when cybernetics has been applied to psychology (Smith and Smith, 1966; Bateson, 1972), management theory (Beer, 1959) or theories of society (Von Foerster, 1953; Mead, 1968; Wiener, 1989). However, what we see in the case of Human-Aided AI is not only transdisciplinarity at the conceptual level – as when theoretical ideas from engineering open up new perspectives in biology, psychology or sociology – but rather a form of transdisciplinarity *at the material level* itself. Networked media that channel human behaviour into data streams that feed into machine learning models, and that, conversely, make machine learning systems feed into human perceptions, feelings and behaviours, do not merely signify a transfer of concepts between academic fields. Rather, those causal loops are materially heterogeneous and hybrid in that they couple social affairs with processes taking place on silicon-based processors. The way news reading, job allocation or dating works today, for instance, is through bidirectional feedback with computational processes relating to predictive AI services.

With predictive analytics in particular, it is important to see that the essential feedback loops that constitute those machine learning systems as Human-Aided AI systems operate at the system level, not at the individual level. While many of these services are described to users as 'personalisation' (personalised ads, personalised search results and the like), personalisation in fact means treating people differently *at scale* to maximise economic benefit.[2] In ad targeting, for example, the cybernetic system is not the single user's experience that is optimised through feedback so that the user sees only relevant ads. Rather, it is the platform and ad service provider that deploy the cybernetic system to 'optimally' target users en masse so as to maximise their advertising revenue. This is why the terminology of 'targeting' fits better than 'personalisation': by analogy with the anti-aircraft firing control system, which is optimised to hit targets *statistically more often*, targeted advertising optimises advertising decisions to engage users *statistically more effectively*. Predictive targeting always means that there is a risk of failure at the individual level; the idea, however, is to look at this risk from an aggregate rather than an individual perspective, and to improve the aggregate performance of the system, even if it still happens that single individuals see ads that are irrelevant to them (see also Chapter 9 on this point).

The aggregate rather than individual logic of the feedback loops involved in predictive analytics makes those systems an important factor – or perhaps we should say facilitator – in structuring our social environment. As feedback loops optimise the differential treatment of *masses* of individuals, their aggregate, system-level impact results in a stratification of society in terms of access to opportunities, possibilities, resources and information. In fact, this form of AI means large-scale cybernetic experiments on societies are already being run by data companies today.

Social inequality, social sorting, racism and sexual exploitation can easily be locked in place and even amplified as an effect of these experiments, as, for instance, Safiya Noble points out in her book *Algorithms of Oppression: How Search Engines Reinforce Racism* (Noble, 2018). To recount one of the autoethnographic case studies she describes, one afternoon in 2011, Noble googled the phrase 'black girls' when she 'searched for activities to entertain my preteen stepdaughter and her cousins of similar age' (p 17). The results on the first pages consisted of sexualised and pornographic content. Consequently, Noble raises the question 'how "hot", "sugary", or any other kind of "black pussy" can surface as the primary representation of Black girls and women on the first page of a Google search' and points out that 'something other than the best, most credible or most reliable information output is driving Google. Of course, Google Search is an advertising company, not a reliable information company' (p 5), as search results are 'actually a reflection of advertising interests' (p 36).

The output of the search engine she cites clearly results from the fact that the system is optimised not for maximum 'intelligence' in providing the individual user with what they're looking for, but for maximum profit at the aggregate level from a business perspective. The entities or 'systems' stabilised by this feedback-driven AI system are thus *the very social structures that provide economic benefit to the operating company*. Racism, sexism, classism and other forms of structural injustice and discrimination are both monetised and co-produced through the AI systems we use every day. These social structures now operate on a new, digital-cybernetic dynamic, with predictive AI as the latest mediator and facilitator of old (as well as new) patterns of injustice. In predictive AI applications, this could also be described as the *performative effect*, where individuals are pre-emptively treated differently based on predicted attributes, which in turn helps to *create* the very social differences the system allegedly predicts.[3] By categorising people into algorithmic groups and acting on those classifications, AI systems don't merely observe or reflect social inequalities, they actively produce and reinforce them. In this way, the cybernetic feedback loop between the aggregate performance of predictive AI and real-world social inequalities *performatively produces* the very divisions it purports to predict.

Pointing this out in critical scholarship requires us, first, to consider AI systems as Human-Aided AI systems, which inherently source their 'intelligence' from human feedback, and second, to regard these Human-Aided AI systems in terms of power arising from cybernetic control – or *control power*, for short. This control power is performative in relation to social structures. This can only be seen in the cybernetic perspective that emphasises the mutual influence between AI apparatuses and society. By thus outlining how Human-Aided AI is the contemporary manifestation of cybernetics, we gain a conceptual tool to reframe AI as a technology of power that inherently produces and stabilises structures of injustice, exploitation and oppression.

Control power: manipulation and discrimination

The essential reliance of Human-Aided AI systems on continued end-user feedback is not surprising if one contextualises the present hype around

AI (the so-called AI summer), not only in the history of (symbolic) AI, but in relation to the history of cybernetics. This fresh look at AI as Human-Aided AI has important ramifications for the conceptualisation of the *power* of contemporary AI apparatuses. In reference to the principle of 'control through feedback', the form of power that emerges from the entangling of AI apparatuses and the social sphere has a strong component of what one could call *control power*. Here I use the term 'control' as synonymous with 'closed-loop control' or the German engineering term *Regelung* (literally 'governing' or 'regulating'), which is opposed to 'open-loop control' or the German term *Steuerung* (literally 'steering'). Closed-loop control is a responsive form of governance that relies on constant monitoring and feedback, while open-loop control is directive, unilateral and prescriptive.

I propose that the manifestation of control power in Human-Aided AI can be analysed along two major dimensions: manipulation and discrimination. First, manipulation refers to the adept wielding of influence over users to control their actions in a specific direction. This is control that operates at the *individual* level of feedback loops, as described before and whose currently most well-known manifestation is nudging (see also Chapter 3). This kind of power generally allows its subjects some leeway to act seemingly in their own fashion and out of free will, but at the same time situates this action within the fine constraints of a personalised environment of algorithmically curated affordances, information, social relations, choices, visual representations and so on to subtly guide it in a specific direction. The automated calculation of such affordances and representations of choices makes use of the predictive modelling of a user's reactions, psychological traits, motivations and impulses. The more fine-grained the ways in which the monitoring and feedback cycles are designed, the more this modus operandi of power can rely on what will be perceived as 'soft' impact instead of prescriptive and repressive interventions.

Second, discrimination is the systematic unequal treatment of different individuals and cases; it makes use of predictive knowledge to pre-emptively differentiate between individuals and cases. This process not only reflects existing social disparities but, through the performative power of AI, *actively contributes* to their production and reinforcement. Discrimination thus always refers to a relation between *many* individuals and aligns with the aggregate, system-level dynamics of feedback processes. As described before, the feedback architecture of commercially driven Human-Aided AI apparatuses is designed to stabilise these structures of inequality, as such disparities can be monetised and turned into profit within the logic of AI business models in networked media. Consequently, the very feedback loops that drive AI systems performatively shape the social realities they engage with, embedding inequalities further into the fabric of society.

This bidimensional conception of control power as it develops from the consideration of the cybernetic feedback principle shows that the risk of manipulation and discrimination of users is *inherent* to contemporary AI. Insofar as contemporary AI is Human-Aided AI, to secure the immense redistribution of wealth from the bottom to the top that is facilitated by contemporary digital business models, those apparatuses must rely on the 'voluntary' and 'pleasurable' participation of many users and therefore can only afford a comparatively soft and non-repressive form of power. This is control power – that is, governance by the principle of closed-loop as opposed to open-loop control.

This is not to say that there are no repressive or restrictive effects of predictive AI. For example, if a loan is denied to a person based on a predictive credit assessment algorithm, this is a restrictive effect. Nevertheless, all these systems generally leave their subjects a controlled amount of leeway for their 'own' choices and failures. If user behaviour were propelled entirely by a predictive AI system, the AI system would not be able to learn any improvements from the data collected about users' choices. Credit scoring services, to stay with this example, are usually large, cross-industry platforms that combine data from many businesses and sectors, so that each individual can be tested and observed multiple times and on different occasions – for example, when not only banks but also phone operators, internet providers and insurance companies submit their customers' data to the same credit scoring service. Hence, the openly restrictive and repressive manifestations of control power are only the tip of the iceberg, which is small in comparison to the vast underpinning of everyday monitoring and nudging that often receives less attention as a form of power.

The cybernetic leviathan?

Wiener himself was worried about the broader societal impact of the diverse manifestations of cybernetics principles in many areas of engineering, economics, politics and society. In his book *The Human Use of Human Beings* (Wiener, 1989), he engages with many objections and anxieties related to the proliferation of cybernetics-based automation in the factory, in military strategy and in politics, confirming many of these worries. Among other implications, he picks up on the discourse of the 1950s about the prospect of a *machine à gouverner*: a cautionary thought experiment promoted in the French debate by the philosopher and theologian Dominique Dubarle (1907–87). Dubarle, one of whose articles Wiener quotes at length, adopts Wiener's notion of the cybernetic 'prediction machine', extending it to the dystopian idea of a *machine à gouverner*: a universal 'State apparatus covering all systems of political decisions' on the planet (cited after Wiener, 1989: 178). On the one hand, Dubarle appeases the reader by affirming the purely theoretical nature of this debate, because 'the *machine à gouverner* is not ready for a very near tomorrow', as the immense 'volume of information to be collected and to be treated rapidly' poses 'very serious problems' that go 'beyond what we can seriously dream of controlling' technologically and mathematically today (pp 179–80). However, at the same time and with reference to political examples of the day, Dubarle contends that with the rapid spread of cybernetics-based automation: '[w]e are running the risk nowadays of a great World State, where deliberate and conscious primitive injustice may be the only possible condition for the statistical happiness of the masses: a world worse than hell for every clear mind' (Wiener, 1989: 180).

Dubarle adds that if this world state, forming a huge *machine à gouverner*, becomes real, we face 'the rise of a prodigious Leviathan. In comparison with this, Hobbes' Leviathan was nothing but a pleasant joke' (p 180).

In the history of political philosophy, Hobbes' theorised figure of the leviathan is a justification of absolute sovereign rule to which citizens voluntarily submit in order to avert the even more hellish 'state of nature'. In this conception of sovereign power, the idea of a *voluntary*, even contractual participation of the citizens who collectively endorse the sovereign regime is central. The world state of the *machine à gouverner* is then, now in Wiener's paraphrasing of Dubarle, 'a great superhuman apparatus working on cybernetic principles' (p 182). While Wiener shares with Dubarle this rather bleak outlook, he adds an important refinement of Dubarle's argument. This refinement counters the idea that the emergence of a planetary cybernetic communication apparatus (*machine à gouverner*) would lead to a new form of *sovereign* power. In Wiener's words:

> [The] *machine à gouverner* [...] is not frightening because of any danger that it may achieve autonomous control over humanity. [...] Its real danger [is that] such machines, though helpless by themselves, may be used by a human being or a block of human beings to increase their control over the rest of the human race or that political leaders may attempt to control their populations [...]. (Wiener, 1989: 180–1)

This insight from the 1950s is not only striking as we witness the impacts of today's networked media on our democracies and societies, but also an apt rebuttal of all the doomsday AI critiques that are currently in vogue. Already 70 years ago, Wiener was articulating quite an accurate description of what is really frightening about AI: not that it will eventually 'take over' humanity, but that it will establish a pervasive control apparatus to the economic and political benefit of a very small number of actors. This control apparatus stabilises and capitalises on societal inequalities and discrimination, hate speech and misinformation, racism and sexism. To quote Dubarle again, the emergence of such a cybernetic machine comes with 'deliberate and conscious primitive injustice' that is the 'condition for the statistical happiness of the masses' (p 180). In contemporary Human-Aided AI, this 'happiness' is barely more than the guilty pleasure that makes us use our networked media services every day.

Opacity in Machine Learning and Predictive Analytics

There are numerous ethical problems associated with the control-type power implemented by predictive AI that relies on feedback according to the Human-Aided AI principle. In the following three chapters, I will briefly discuss three of them: the issues of explainability, unfair bias and collective responsibility. In my discussions, my main focus is on showing what the analysis of predictive systems as cybernetic Human-Aided AI systems can contribute in addressing these ethical issues.

An important ethical debate relating to machine learning in general and predictive modelling in particular revolves around opacity, which can also be phrased as lack of explainability. Jenna Burrell pinpoints the problem in a much-cited paper from 2016: '[These algorithms] are opaque in the sense that if one is a recipient of the output of the algorithm (the classification decision), rarely does one have any concrete sense of how or why a particular classification has been arrived at from inputs' (Burrell, 2016: 1; see also Matthias, 2004; Mittelstadt et al, 2016). It has since become a common trope in the ethics of AI that the ethical use of automated decision-making systems requires the ability to provide reasons to the people about whom decisions are made. Coeckelbergh (2020b), for instance, stresses that the ability to provide reasons and explanations is a key component of moral responsibility, which he interprets as 'answerability', on the part of the operator of the automated decision-making system towards moral patients – that is, those individuals affected by the system.

Before we examine what the cybernetic perspective of Human-Aided AI can add to this debate, it is useful to revisit how Burrell teases out the problem of opacity into three different strands:

1. '[O]pacity as intentional corporate or institutional self-protection and concealment' (Burrell, 2016: 1–2). This includes cases where business principles or the rationales behind decisions are kept as corporate secrets or are misrepresented as part of a deceptive manoeuvre.

2. Opacity due to widespread technical illiteracy – for example, the inability of most non-specialists, including the general public, to read and understand code.

3. Opacity due to certain fundamental properties of machine learning algorithms: 'opacity that stems from the mismatch between mathematical optimization in high-dimensionality characteristic of machine learning and the demands of human-scale reasoning and styles of semantic interpretation' (Burrell, 2016: 2).

A common epithet for diagnosing opacity is the term 'black box', introduced into the discourse most notably by Pasquale (2016), who was largely referring to intentional opacity (type 1), although nowadays most people tend to associate 'black box' with the machine learning–specific type 3 opacity. Type 1 opacity is really a kind of *unwillingness* to explain on the part of the explainer, because it is purely strategic and capitalist in nature (opacity is intentionally created by withholding information from competitors, regulators and the public as a matter of corporate strategy). Type 2, on the other hand, relates to the limited capacities and technical understanding of the recipient of explanations, which is in part an issue of education and training and a consequence of the division of labour. Type 3 opacity, then, seems to be the core issue that really points to the lack of *explainability*. Type 3 has fuelled the emergence of the research and engineering field of 'explainable AI' (XAI) as a subfield of machine learning, which develops methods and procedures that allegedly help operators, users and affected persons in 'understanding, visualizing and interpreting' how deep learning and other machine learning models arrive at their outputs (Samek et al, 2017).

In the discourse on XAI, especially with regard to type 3 opacity, AI's principal lack of explainability is sometimes rendered as a natural tribute we have to pay when we invoke the levels of intelligence some modern machine learning systems perform when they 'excel in a number of complex tasks' and 'can even outplay professional human players in difficult strategic games such as Go' (Samek et al, 2017: 1). As an example, Samet et al mention '[t]he 37th move in the second game of the historic Go match between Lee Sedol, a top Go player, and AlphaGo, an artificial intelligence system built by DeepMind' (p 1):

> AlphaGo played a move which was totally unexpected and which was commented on by a Go expert in the following way: 'It's not a human move. I've never seen a human play this move' (Fan Hui, 2016). Although during the match it was unclear why the system played this move, it was the deciding move for AlphaGo to win the game. (Samek et al, 2017: 1)

While in AlphaGo this opacity might be morally irrelevant, so the paper continues, in other applications of AI, such as medical diagnosis and self-driving cars, 'it would be irresponsible to trust predictions of a black box system by default' (p 1). Developing technical 'methods which help to better understand what the model has learned' and 'for explaining individual predictions' is therefore necessary for the verification, improvement and legal compliance of AI systems, but would also enable 'learning from the system'.

In this framing of the issue of explainability of predictive AI systems, there is a latent hypostatising of AI systems as artificial 'geniuses' whose exceptional and original insights we can tap into at the cost of perhaps being unable to understand them. Explaining the genius' decision is then an almost forensic (and necessarily retrograde) activity, a slow and painstaking process, and with a focus on verification of and learning from the machine's apparently superior and faster 'reasoning'. In opposing and deconstructing this tacit vision of genius AI, I will argue that in most cases, unexplainable predictions, classifications or decisions by AI systems are not a result of any exceptional insight or creativity exhibited by the AI system itself. Unexplainability is usually the result of an aggregated and statistical form of reasoning that lies behind the design and development of AI systems that are deployed to further the economic and profit-oriented goals of their operators.

The demise of explainability

For this argument, it is unnecessary to dive deeper into why explainability is *technically* such a hard problem in the case of machine learning models. Rather, we shall take a step back and look at the consequences of the economic reality of predictive systems as cybernetic, Human-Aided AI systems. My argument shifts the focus away from the question of why explainability is technically difficult or impossible, even though it is ethically important, to the more power-analytic and critical question of why creators and operators of such systems, following a cybernetic mindset, are often incentivised to ignore this issue.

In the way that the 'intelligence' of cybernetic AI systems lies in the design of feedback loops (see Chapter 2 and the second thesis of Human-Aided AI), predictive AI systems follow a purely functional (non-hermeneutic) view of the behaviour they are designed to predict. 'Giving reasons' is then, both technically and economically, pretty much an obsolete undertaking. For example, the control system for anti-aircraft fire does not strive to understand *why* a particular aeroplane is making this or that unexpected turn; it is only concerned with computing an economical 'best bet' with regard to the future trajectories of both plane and missile based on what the system has been able to 'learn' from the trajectories of many previous aircraft in similar situations (see Chapters 7 and 8). The same principle of economics applies for predictive systems that target users as potential customers, insurance clients, job applicants, dating matches, security risks and so on. These systems do not model people in terms of their motives, identities or possible volitional decisions. They instead compute for each case a statistical 'best bet' on how to treat (target, score, classify, predict) that case. This computation is based on comparing the information available about the case (auxiliary data, which correspond to the past trajectory of the targeted aeroplane in the example of anti-aircraft gun control) with recorded performance data from millions of past cases (training data, which correspond to the recorded trajectories and successful or unsuccessful attacks on many previous aeroplanes).

The rationality behind the design of these systems does not deal with the 'reasons' governing how cases or individuals are to be treated based on some model of their motivations, goals, volitions and the like. The mechanism that produces the targeting decision is a statistical and behavioural one that is based on the history of cases to which the present case is compared (see Chapter 5). Predictive AI systems are not designed to find out 'the truth' about or 'do justice' to the individuals to whom their predictions are applied. Instead, these systems are designed to treat *masses* of cases most effectively from the *aggregate* point of view of the operators, who are trying to maximise certain aggregate (or average) goals, such as overall profit or general safety. Crucially, in this rationale, operators are willing to take the risk of treating some individual cases unfairly along the way, because this will be rational from an aggregate perspective when the predictive system still produces a profit margin. The criterion that is implicitly or explicitly in place to optimise the algorithmic systems in question is that they *perform well* economically, and that means to perform *sufficiently well* on a *sufficient number* of cases, rather than to discover the 'truth' of every single case.

Predictive performance as the new truth

So, as a consequence of the economic reality of predictive AI systems as cybernetic, Human-Aided AI systems, we can see that these systems are not designed to align with what one might call 'the truth about' or 'doing justice to' each single case, but with overarching performance goals. That is, predictive systems are designed to compute a *best bet* on how to treat each specific individual so as to produce an overall performance margin. The epistemic and universal category of 'truth' is in predictive analytics replaced by a pragmatic and economic measure that I refer to as 'predictive performance'. In Chapter 7, a variant of this insight was already debated when Leo Breiman was advocating for a shift towards 'predictive accuracy' as a criterion of algorithmic modelling. In this subsection we will now push this analysis further, replacing 'predictive accuracy' with the more accurate term 'predictive performance' to highlight the key role of economic rather than epistemic criteria for 'accuracy' in the design of cybernetic predictive AI systems. To sharpen this analysis, I will take a slightly more formal approach in providing a schematic definition of predictive performance that serves as a tool to illustrate some of the consequences of the cybernetic nature of these AI systems.

Imagine that an insurance provider deploys a predictive model A as a new risk assessment mechanism to decide for each insurance claimant $i \in I$ whether the insurance provider should accept $(A_i = 1)$ or reject $(A_i = 0)$ them. Let's assume the insurance company has a baseline assessment mechanism B in place, which is just the state-of-the-art decision model that the company has been using so far in making their risk assessment. The idea is to develop A so that it performs better than B. Let $B_i \in 0,1$ denote the output of this old decision mechanism for claimant i. Moreover, let V_i denote the real case-based profit associated with the case of claimant i (total gain or loss) if i were to be accepted as a customer. Now, it seems economically 'reasonable' – in the absence of external ethical or regulatory constraints – to optimise model A in such a way as to maximise the following quantity, which I call the (relative) 'predictive performance':

$$\Delta P = \sum_{i \in I} \left(A_i - B_i\right)V_i$$

This formula measures the predictive performance of the new assessment mechanism A as the difference in total profit (the profit margin ΔP) compared to the baseline assessment mechanism B.[1] A formula like this provides insight into the kind of feedback loops that need to be implemented to help the model converge – through constant retraining and recalibration – towards the performance goal. The goal is to maximise ΔP (the profit margin resulting from the deployment of the new decision mechanism); the feedback loop that supports this goal is based on a retrospective evaluation of the case-based profit V_i.

In this feedback loop, cases $i \in I$ with positive $\left(A_i - B_i\right)V_i$ reinforce the decision mechanism, while cases with negative $\left(A_i - B_i\right)V_i$ lead to a penalty in the form of an incremental correction of the assessment. As encoded in the formula, positive reinforcement occurs whenever the new model improves on the baseline decision mechanism by either newly admitting a case with positive profit V_i or newly rejecting a case with negative profit V_i. Negative 'punishment', on the other hand, happens only when the new model makes mistakes *that did not occur in the old assessment* by either newly admitting a non-profitable case or newly rejecting a profitable one.

Note that cases (profitable or not) where the new and the baseline assessments match do *not* result in feedback. The optimisation procedure is like the thermostat in the previous section: it considers only differences to the status quo, not absolute values. These arguments highlight two important points to consider in the economic use of predictive models in the absence of external constraints or regulations:

Lack of attention to marginalised groups: Cases or groups of cases with a higher potential profit margin V_i (either for the individual case or due to the large size of a group of similar cases) carry more weight in the optimisation of the model. This explains why marginal cases – marginal both in terms of individual weight and group size – are less likely to make an impact in the feedback-based optimisation process of the assessment model.

Biases will only be corrected if that brings extra profit: The perpetuation of biases that already exist in the baseline method has no negative weight in the design of feedback loops as reflected in my formula. The correction of biases is only evaluated positively if it also produces profit for the company. This reflects the fact that we have little reason to expect that companies will, without external constraints, optimise for antidiscrimination if this does not produce an additional positive profit margin. Regulation must therefore find mechanisms to adequately factor in discrimination as negative profits in the companies' business cases.

Case-based vs system-level explainability

In the use of predictive AI systems, there is an evident shift in epistemology from finding some truth about individual cases towards efficiently handling masses of cases so as to produce an aggregate performance margin. This observation has consequences for the problem of opacity in predictive modelling. For a start, explainability of the model's output on a case-specific input is indeed lacking if the training of the predictive model does not rely on any case-based feedback that reflects human-intelligible assessments of each single case. Looking at the decision-making routine from a case-based perspective, the model *is* opaque in the sense of type 3 opacity. But this opacity seems to result less from the fundamental 'mismatch between mathematical procedures of machine learning algorithms and human styles of semantic interpretation' (Burrell, 2016: 3), and primarily from the fundamental mismatch between what should be a value-driven and socially intelligible assessment of an individual case, but what is in reality the result of optimisation for aggregate goals. If the operator is not interested in treating cases fairly or on factual grounds that can be argued for each single case, why would fair treatment matter to a decision-making routine optimised towards aggregate goals that align with such case-based reasoning? The lack of explainability in a *case-based perspective* is thus certainly not a mathematical limitation, but a consequence of how (and which) mathematics is used to implement a system embedded in an economic context where that system is optimised towards achieving certain performance goals.

In demanding explainability, we could, however, switch to a more systemic perspective as a way of explaining that really reflects the economic reality of such systems. This would mean turning towards the feedback loops that drive the training and retraining of the decision model, which are, according to the Human-Aided AI paradigm, a constitutive part of the AI system. The architecture of these feedback loops *is certainly explainable* and not at all opaque, at least not for technical reasons. This architecture is a direct manifestation of the business model behind the deployment of the predictive system, and detailed information about it could be very helpful in *disclosing and explaining* the functioning of the system to those who are affected by its decisions and to the public at large, if this information were not usually guarded as a trade secret. Explaining the feedback loops would transform type 3 opacity, which pertains to the case-based perspective, into a problem of type 1 opacity, which is purely political in nature.

Hence, operators, when acting in an 'economically rational' manner, *do* employ the predictive system in an accountable way, but accountability when viewed through the lens of their (cybernetic) rationality does not mean 'doing justice' to the individual case, and does not imply the ability to explain every decision at a per-case level, but rather to detail how exactly the assessment system has been optimised towards business interests that adhere to an aggregate perspective. The demand for case-based explainability, in short, confronts an economically driven system with the subject's demand for individual justice and truthfulness. This mismatch is thus not primarily between the technical properties of machine learning models and the cognitive workings of human semantic reasoning.[2] More profoundly, the mismatch concerns the idea of cased-based explainability, which (sadly!) is practically irrelevant to the cybereconomic reality in which these models are interwoven with business goals via feedback loops.

Two demands of explainability

Following this distinction between case-based and system-level explainability, we can articulate two escalating ethical and regulatory demands of explainability:

Demanding system-level explainability: Much would already be gained by compelling operators to explain the feedback loops of AI systems – that is, to disclose what data they use to retrain and calibrate their system, in what precise weights and towards which goals. This amounts to disclosing the real version of the speculative (and likely simplified) formula for predictive performance I gave previously, alongside detailed information on what feedback data are used and how the different steps in training the system and 'fitting' the model to the training data translate into explanations that are accessible to human semantic reasoning. Calling for this form of system-level explainability is a technically simple and feasible demand, since the principal (type 3) opacity applies to the case-based level, whereas from the holistic, aggregate perspective, those systems are opaque only in terms of type 1 (obfuscation and corporate secrets).

Disclosing relevant corporate secrets by granting access to source codes and engineering decisions is often debated in terms of transparency, which in turn is often associated with auditing and oversight (see also Burrell, 2016: 10). In contrast to this approach, I argue here for a strategy of explanation that enables 'answerability' (Coeckelbergh, 2020b) to the moral patients of predictive AI systems. My claim is that explaining feedback loops *is* a valuable and human-accessible explanation of the cause–effect chains that govern the behaviour of predictive systems as they appear from a systemic perspective. If we acknowledge the cybernetic logic of machine learning technology, its behaviour is as explainable as it gets. This is not to say that predictive modelling is ethically unobjectionable (because, allegedly, the problem of opacity is now 'solved'). Quite the opposite – only by recognising this cybernetic logic can we fully engage in an ethical debate about the true stakes of adopting predictive AI.

What is ethically troubling about these systems is not just their opacity, but their efficiency-based, rather than truth-, justice- and human-centred rationality. Due to their focus on the aggregate level, these systems have *performative effects* on social structures, actively creating and reinforcing social differences rather than merely reflecting them (Chapter 8). The concern is not only the lack of transparency, but the fact that we live in a society and under a form of capitalism in which the single individual *does not count* because what matters is the efficient management of masses and swarms of individuals – whose treatment is dictated by economic optimisation rather than ethical considerations. On the one hand, therefore, demands for explainability should focus on system-level explainability, specifically the *explainability of feedback loops*. This entails explaining not just individual case decisions, but the aggregate economic rationale of the device as it is dynamically embedded in society and economy.

Demanding case-based explainability: On the other hand, my argument shows that imposing case-based explainability is a much stronger and potentially more far-reaching ethical and political instrument against the risk of such systems becoming unhinged in a completely unregulated space of cybernetic reasoning. While the system-level explainability of feedback loops is purely descriptive (you don't need to alter the system, 'just' lift corporate secrets), case-based explainability will not be successful as a descriptive, forensic approach (unless AI is hypostatised as genius AI, as in the earlier discussion of AlphaGo). Rather, imposing case-based explainability will entail demands that reach deeply into the technological layer, as the research on XAI shows. An entirely new scheme of evaluating, monitoring and reasoning would be imposed onto the design and operation of such systems, demanding human semantic interpretability of the model-level mathematical operations that handle the high dimensionality of machine learning problems. This would amount to introducing a philosophical and epistemological regime of 'giving reasons' and 'doing justice to the individual case' that is simply not provided for in the statistical logics of feedback-based aggregate profit maximisation.

Ethics at the end of theory

The preceding discussion about the opacity vs explainability of predictive risk scoring systems is connected to a much broader and intriguing debate in the philosophy of science and digital humanities, which deserves brief mention. The narrative around data analytics and AI often evokes the idea that truth can be 'found' in the world through the collection of sufficient amounts of data, suggesting that 'objective' truth is out there, waiting to be 'discovered'. For example, as Rob Kitchin (2014) recounts in his critical essay on the proclaimed 'end of theory' (see Anderson, 2008; see also Chapter 7), some proponents in the digital humanities believe that data-driven techniques 'bring methodological rigour and objectivity to disciplines that heretofore have been unsystematic and random in their focus and approach' (Kitchin, 2014: 7–8). It is suggested that data-based methods will finally bring objectivity even to disciplines that have traditionally been difficult to free from subjectivity and arbitrariness.

Anderson's bold claim, that obtaining 'objective' results through AI methods is merely a matter of amassing sufficient quantities of data, was met with a striking refutation by two mathematicians: 'Using classical results from ergodic theory, Ramsey theory and algorithmic information theory, we [...] prove that very large databases *have* to contain arbitrary correlations. These correlations appear only due to the size, not the nature, of the data' (Calude and Longo, 2017: 595, italics added).

In philosophy, the belief that data will help us uncover objective truth is a stance that philosopher Antoinette Rouvroy, a scholar of algorithmic governmentality, calls a 'naive realism' (Rouvroy et al, 2022: 125). It neglects the fact that truth is not 'found' – not even in the world of data-driven decision-making and scientific inquiry. For some, the field of digital humanities seems to represent a revival of the flawed 'culture of objectivity' (Traweek, 1988: 162; see also Haraway, 1988: 581), a digitalised version of 'the god trick of seeing everything from nowhere' through data (Haraway, 1988: 581). As Orit Halpern et al have pointed out: '[I]n our era, data is not simply descriptive or analytical but actively constructive. The assumption that complex datasets yield the most comprehensive truth returns us to the problem of theory and a question of history' (Halpern et al, 2022).

This 'question of history' leads us to Foucault, who, like others, pointed out in general terms that truth is *produced* within socially and historically contingent power-knowledge regimes. By implication, the collection and analysis of data are also situated within such regimes. As others have noted more specifically, *there is no raw and objective data* (boyd and Crawford, 2012; Gitelman, 2013). Instead, we must examine the 'inherent politics' pervading the collection and analysis of datasets (Kitchin, 2014; see also Matzner, 2024: 123–30).

This politics is not solely the result of intentional human decisions. Rather, following the insight from Chapter 8 that predictions *create* reality and AI systems have performative effects, the power effects of data, and their use in knowledge production, create a dynamic by which economic interests find their immediate, technological agency. Given the cultural diagnosis of an ongoing shift towards predictive performance as the 'new truth' criterion in the social, technical and economic assemblages of AI, we must concede that the end of theory might be closer than we think. Today, a person's future is more likely to be treated according to the premises of correlation than based on any profound understanding of their skills, character or interests.

With the proliferation of AI technology across so many domains, reality is already enacting the end of theory. If knowledge is always situated within power apparatuses (Foucault), and if the sociotechnical assemblages of Human-Aided AI *are* power apparatuses, then this constellation will potentially shift the dominant forms of knowledge and indeed the very meaning of 'truth'. The claim that the new, big-data methods are actually flawed is certainly 'true', but true in the *old* understanding of truth and theory.

Similar to the two demands of explainability, we can also formulate two ethical demands concerning 'theory'. The first, weaker one refers to the aggregate level, where the entanglement of knowledge production with large-scale economic interests must be explained. The second and stronger demand is for 'proper', 'good old' theory based on causal modelling, at least when it comes to producing knowledge about individual cases that determines individual fates.

10

Bias in Cybernetic AI Systems

While demands for AI explainability mostly refer to case-based explainability in the popular understanding, there is also an ethical debate that takes a more aggregate perspective with regard to decisions, classifications, texts and images as outputs of AI systems – the debate around bias in AI. The claim of bias is always a claim about a systematic pattern in the outputs of an AI system that is only evident when comparing many different cases. Among the hundreds of examples of the racial, gender and class biases of AI systems are studies that show how women are less likely to be shown ads for highly paid jobs by ad-targeting AI systems (Gibbs, 2015); that search engine results (both text and images) boldly reproduce gender and racial stereotypes (Noble, 2018); that hiring algorithms (Raghavan and Barocas, 2019) as well as AI translation tools (Savoldi et al, 2021) tend to exhibit racial and gender biases; that a widely used health management system in the US that scores the relative urgency of medical cases was found to be severely racially biased Vartan (2019); that face-detection AI systems were found to be less likely to detect in images the faces of women, Black people and, most seriously, Black women (Buolamwini and Gebru, 2018); and that generative AI systems such as ChatGPT routinely reproduce gender, racial and Eurocentric stereotypes (Busker et al, 2023).

From a conceptual perspective, Friedman and Nissenbaum famously defined the term bias 'to refer to computer systems that *systematically* and *unfairly discriminate* against certain individuals or groups of individuals in favor of others' (Friedman and Nissenbaum, 1996: 332, italics in original). With respect to AI, the questions of what exactly is meant by 'unfair' and which 'groups of individuals' are to be considered greatly need clarification. For instance, it is not plausible to assume that unfair discrimination by AI systems always occurs along 'classic' protected attributes such as ethnicity, class or gender; it is entirely possible that AI-based targeting, scoring and classification follow sophisticated and evasive patterns of unfair discrimination that do not map directly onto existing social, let alone legally protected, categories.[1] Important contributions towards an understanding of this impasse come from the theory of intersectionality, which emphasises that combinations of different discriminated-against attributes can result in types of prejudicial treatment that are more harmful and less visible than cumulative single-attribute discriminations (Crenshaw, 1989, 1991; Collins, 2015; Crenshaw, 2015; Buolamwini and Gebru, 2018). Most importantly, intersectionality teaches us that the *epistemic work* of making biases visible and naming them is a non-trivial part of ethics of AI scholarship (see also the Introduction).

It is not the task of this section to engage in the extensive scholarly debate around bias in AI that has emerged in recent years. My goal is to point out an important consequence of the cybernetic theory of predictive AI systems for the analysis of bias. As I will argue, the cybernetic perspective makes visible how biases in predictions can *dynamically stabilise* in feedback loops between the machine learning model and the social reality of an AI system. This only becomes visible when we consider the entire embedded system, not just the model. That is, we must view the machine learning model as being coupled to our social reality via feedback loops from which the system draws its stability through continuous retraining while simultaneously acting on the world. In this way, we can bring to light one specific mechanism of how bias emerges, which I call the 'biased feasibility' of feedback loops (Mühlhoff, 2021). By this term I refer to situations where the design of feedback loops favours more privileged groups or individuals to the detriment of more vulnerable, protected or underprivileged groups.

One example of such a situation is the racialising bias of the Correctional Offender Management Profiling for Alternative Sanctions (COMPAS) system, a commercial risk assessment tool used to predict recidivism in the US criminal justice system. COMPAS is 'used throughout the [US] criminal justice system' (Rudin et al, 2020) to aid in parole and sentencing decisions, and it relies on a proprietary algorithm created by a company formerly called Northpointe, rebranded in 2017 to equivant (Dressel and Farid, 2018). The system has been used on more than a million people since 1998 and has offered its recidivism-prediction module since 2000. According to an investigation by ProPublica, this module has proven to be racist in different ways (Angwin et al, 2016), one of which concerns the distribution of predictive *errors* between Black and White defendants. In a sample of more than 7,000 individuals arrested in Broward County, Florida, in 2013 and 2014, the rate of false negatives (people who were rated as low risk by the system but who actually reoffended) was quite high for White and low for Black defendants, while the rate of false positives (people who scored as high risk but actually did not reoffend) was high for Black and low for White defendants (see Table 10.1; Angwin et al, 2016; Larson et al, 2016).

Thus, the COMPAS system is racially biased with respect to the *risk of prediction errors*. The different kinds of prediction error (false positives vs false negatives) are unequally distributed to the benefit of White and to the

Table 10.1: Assessment of the COMPAS recidivism score on a population of 7,214 defendants in Larson et al (2016)

	Overall recidivism			Violent recidivism		
	All	**Black**	**White**	**All**	**Black**	**White**
False positive rate	32.35	44.85	23.45	27.93	38.14	18.46
False negative rate	37.40	27.99	47.72	47.15	38.37	62.62

Source: Values from Larson et al (2016)

detriment of Black defendants (as the false positive rate is approximately twice, or for violent recidivism three times as high for Black than for White defendants, while the opposite is found for the false negative rate; see Table 10.1). So the risk of being wrongfully imprisoned is disproportionately borne by the group of Black defendants, while the chance of being wrongfully released is much higher for White defendants. This situation of unevenly distributed risk of error points to what I henceforth refer to as 'second-degree bias', which is worth distinguishing from 'first-degree bias' (see also Mühlhoff, 2021).

First- vs second-degree bias

A first-degree bias against a group of Black people (in the previous example of the COMPAS system) occurs when the tool scores Black people as having higher risks of recidivism than White people – something the COMPAS system really *does* do according to Angwin et al (2016). The reasons for this are deeply rooted in the racialised concept of recidivism risks, because drivers for reoffending include a person's socioeconomic position, history of traumas and experiences of discrimination, educational level, neighbourhood, employment and social status. The racial bias inherent in this notion of recidivism risk falls within the scheme of 'proxy discrimination', as described by Schwarcz and Prince (2020): the COMPAS recidivism-prediction tool has the primary goal of discriminating against defendants who are more likely to reoffend. Although the system does not include race as an explicit parameter, its risk scoring seems to correlate with race. Discriminating against defendants who are more likely to reoffend is, in the terminology of Schwarcz and Prince, the 'discriminator's facially neutral goal' of the COMPAS system. If, their analysis continues, 'membership in a protected class [here, Black people] is predictive of a discriminator's facially neutral goal', this 'mak[es] discrimination "rational"' (Schwarcz and Prince, 2020: 1257). This is the essence of 'proxy discrimination', in which a supposedly uncontroversial discriminatory goal correlates with, but is not caused by, membership in a protected class.

I refer to such outcomes as instances of a first-order racial bias. Calling this a form of bias rather than 'rational' discrimination unmasks the hidden racism in the purpose for which such systems are put in place. This bias is already implicit in the idea of predicting criminal recidivism and in the very purpose of the AI system that was built upon that idea. First-degree bias differs from second-degree bias in that it's not the prediction *error* that is considered, but the classification itself, insofar as it is taken to be 'non-erroneous', 'facially neutral' or 'rational' (see also Schwarcz and Prince, 2020). First-order bias is thus not a 'flaw' in the model, but is purely social in origin, being deeply rooted in the intentions of the creators and users of these systems and in the episteme from which these intentions are judged as valid.[2] Because of the inherent racism behind the idea of predicting recidivism or criminal risk, there are fundamental moral reasons to 'abolish' the use of AI systems for sentencing decisions (Schwerzmann, 2021). I fully agree with these judgements. COMPAS is first-degree biased, meaning that the very intentions for adopting this system are already (racially) biased, which to my mind is reason enough to abolish the project.

Despite this, however, I propose in this section to draw attention to the concept of *second-degree* bias, of which COMPAS is again a cautionary example. As pointed out earlier, while first-degree bias is always social in origin, only second-degree bias *could* be assumed to be technical, as it seems to be related to flaws in the model (as they are prediction *errors*, after all). As I will now discuss, this intuition is false, as *not even* second-degree bias can generally be resolved by purely technical measures. For this argument, the cybernetic perspective is particularly useful.

Biased feasibility of feedback loops

Let us now ask the purely theoretical question of *how* COMPAS could be modified to control and minimise the risk of false positives and negatives (which, to reiterate, would not make it an ethically acceptable tool, as this would only solve second- and not first-degree bias). Controlling the risk of false positives and negatives would require feedback processes that monitor the actual performance of the system. That is, continuous data collection must occur with regard to whether the people to whom the model is applied actually reoffend, and these data must be compared to the system's prediction in order to continuously retrain the system. However, collecting these feedback data is much easier for false negatives (wrongfully released people, who are more frequently White) because they come back into contact with the criminal justice system after reoffending, and thus feedback data on them will be available. In contrast, it is difficult to impossible to detect false positives (wrongfully incarcerated persons, who are more frequently Black) because it can hardly be determined that a person who remains imprisoned has (not) reoffended. The racial bias of the system is thus rooted in a skewed *principal feasibility* of the necessary feedback loops.[3]

This situation is typical of many socially embedded AI systems – and AI systems, this book argues, are always socially embedded. In predictive hiring, where a machine learning model predicts the work performance of job applicants,[4] there will be more data on the job performance of admitted applicants that can be used *ex post* to validate the prediction and reduce the risk of false positives (wrong hires) than there are data on rejected applicants. Predictive policing will lead to more minor offences recorded in those groups or geographical areas that are under increased surveillance, which would otherwise evade police reporting, thus skewing any *ex post* verification of the predictive system. And dating sites that facilitate contact only between algorithmically 'matched' pairs of users have almost no means of detecting false negatives (pairs of users who are erroneously not brought in contact), which makes it very hard for those systems to wipe out racist, sexist, homophobic and transphobic biases in the matching algorithm that suggests potential dating partners.[5]

The problem of second-degree bias – that is, the biased principal feasibility of feedback loops – primarily concerns the availability of feedback *data* on the actual outcome of the predicted attribute, as the previous examples illustrate. Adding to this, we might also suspect that the people and institutions in charge of building and operating these systems may have strongly asymmetric interests in controlling for false positives and false negatives. For example, development and funding decisions regarding the COMPAS system may be predominantly driven by White interests that effectively prioritise the validity of the tool for White rather than Black defendants. And in predictive hiring – especially in the low-wage sector, where such tools are most commonly employed – the costs of a wrong hire (false positive) to the operator are much higher than the costs of a missed hire (false negative). This suggests that, beyond the skewed availability of feedback data, there may also be second- and higher-order feedback loops that unfairly amplify the already biased distribution of the risk of erroneous treatment that is the inevitable by-product of predictive rationality.

11

Collective Responsibility in the Ethics of AI

From Human-Aided AI apparatuses in general to predictive AI as a dominant application, one common feature clearly emerges: we are all implicated in what AI technology is and does today. When considering how to protect ourselves, others or society as a whole from the risks of this technology, simply rejecting technology or abstaining from digital services are hardly viable solutions – not unless we *all* do so. Even then, such a course would not be desirable, as many contemporary digital services and networked technologies provide significant benefits to our daily lives that we would otherwise lose.

Overall, it seems that the collective enabling of contemporary AI, particularly as Human-Aided AI apparatuses, presents a constellation that is largely overlooked in ethical discussions. A defining feature of this constellation is that the data-generating behaviour of some individuals has consequences for the entire community. This behaviour, as a behaviour that is shared by many (though not all) members of society, holds significant moral and political relevance. Much of modern ethical discourse in the Western philosophical tradition, however, focuses on the moral actions of individual agents. Indeed, most of Western ethics is a discourse that assumes that there *is* a moral agent who, *as a singular subject*, is amenable to ethical discourse and reasoning with respect to their ethical decision-making. But no single one of us is or can be held morally responsible for the fact that platform companies have the power to produce predictive models based on our aggregated data.

When it comes to the ethics of AI, we are in a situation where we need to open up a level of debate beyond the classical ethical question of what is the right action to be taken by a morally responsible individual actor. When we address users' inevitable collaboration in and with AI systems, focusing on individual moral agency and blameworthiness is not a fruitful starting point for ethical discussion. The availability of sufficient data to produce these societal effects is the result of a crowd phenomenon, in which all of us, as users, participate by carelessly exposing our data to platforms (see Kear, 2022). Crowds, however, are not accessible as moral subjects to classical ethical reasoning – and even if some individuals in the crowd engage in moral deliberation, they cannot necessarily alter the crowd's collective behaviour. This constellation, much like the climate crisis, is an instance of what could be termed 'collective responsibility' (see also Young, 2011). In both contexts, no single individual is solely responsible; instead, responsibility rests with all of us as a social and political community, insofar as we (or many of us) share certain behaviours that contribute to risks and harms for others or for all of us.

Responsibility for potential harms resulting from our shared data practices is collective in a way that goes beyond the concept of 'distributed responsibility', which is sometimes referenced in AI ethics when addressing the problem of attributing responsibility for, say, the actions of an automated system (Floridi, 2016; see also Taddeo and Floridi, 2018). Distributed responsibility, applied in the present context, would suggest that each everyday user, as a data provider, is part of a larger network of individualised moral actors – such as designers, programmers, business strategists, investors – each contributing a small but *cumulative* part of the overall action. However, the idea of causal additivity or decomposability of the total action is not a plausible theory in the present context. The users' shared behaviour in sum *enables*, but does not cumulatively *cause*, the creation of predictive AI. Moreover, this enabling does not stem from distinct microcontributions from each individual user, but from the behaviour of all users as a (widely) *shared* grouping. That is, the enabling function of the users' shared behaviour is not simply added together from the causal contributions of individual users; rather, it resembles the dynamics of a crowd, which is a resonant and feedback-driven phenomenon in the sense of cybernetic control power, wherein the whole is greater than the sum of its parts.[1]

Because of the collective nature of this ethically relevant behaviour, it makes no sense to single out the particular responsibility of any specific user or to suggest that, by *avoiding* data-driven services, an individual user could prevent predictive analytics or evade their own co-responsibility. In fact, even the behaviour of deliberate non-users, as an inverse behaviour, is often a response to the shared behaviour of most users and is therefore part of the collective behavioural disposition that is morally at stake. Collective responsibility in the present context refers to the overarching fact that a widely shared behaviour in our society makes certain risks and harms possible. This shared behaviour is a feature of our collective social, cultural and economic landscape, rather than the product of moral actions by isolated individuals.

Hence, the purpose of introducing the concept of collective responsibility in the present case is not to *attribute* responsibility in a backward-looking sense, as in looking for antecedent behaviours – which would amount to an essentially forensic exercise of assigning individual shares of moral blameworthiness to those involved in causing a specific harm. Collective responsibility in the present context is, rather, a forward-looking issue: it entails that we, as a social and political community, *take* responsibility and perhaps even *form* that very community around this emergent sense of responsibility, because this issue pertains to the moral conduct of all of us together as a society.

Iris Marion Young's concept of 'collective responsibility' comes closest to the forward-looking conception I am aiming at here. She employs collective responsibility alongside the notions of structural injustice and political responsibility when referring to constellations wherein 'injustice occurs as a consequence of many individuals and institutions acting to pursue their particular goals and interests, for the most part within the limits of accepted rules and norms' (Young, 2011: 52). Using housing discrimination as an example, she describes how the actions of various players – such as landlords, real estate agents, tenants and buyers – combine to create an unjust set of conditions that make housing inaccessible to poorer sections of society. However, addressing the emergent effects of this constellation as a collective, rather than individual, moral wrong is not straightforward for the liberal mindset deeply ingrained in Western societies and ethics, because in these societies '[t]he attitudes and behavior of the majority of people is so privatized that there exists little organized public space in which actors can appear to others with their judgments of events, let alone join in collective action to transform them' (Young, 2011: 86).

This, according to Young, prompts a sense of collective responsibility, which she frames as *political* responsibility, following Hannah Arendt (Arendt, 1963, 1987). Young argues that this concept of political responsibility is inherently forward-looking:

> I see something essentially forward-looking about this idea of political responsibility. This responsibility falls on members of a society by virtue of the fact that they are aware moral agents who ought not to be indifferent to the fate of others and the danger that states and other organized institutions often pose to some people. This responsibility is largely unavoidable in the modern world, because we participate in and usually benefit from the operation of these institutions. The meaning of political responsibility is forward-looking. One *has* the responsibility always *now*, in relation to current events and in relation to their future consequences. We are in a condition of having such political responsibility, and the fact of having it implies an imperative to *take* political responsibility. (Young, 2011: 92)

It fits our present context that, in Young's description, collective responsibility becomes political the moment it is understood as proactive caring for the future and for 'the fate of others' (Young, 2011: 92), insofar as their fate may be affected by our collective behaviour. It is 'our' behaviour because the focus is not solely on the individual's behaviour, but on that of all of us as members of a political community.

In another respect, however, Young's account requires slight adjustments to reflect the present case. This relates to her apparent focus on political, societal and state *institutions*. Similar to Arendt's reasoning (especially Arendt, 1963), Young focuses on cases where '[t]he imperative of political responsibility consists in watching these institutions, monitoring their effects to make sure that they are not grossly harmful' (Young, 2011: 88). Is widely shared data-exposing behaviour an institution? Potentially yes, depending on what is meant by 'institution', but it is certainly not in Arendt's sense and not clearly enough in Young's sense. In contrast to the state-focused perspective that underpins Young's account, the present context involves quasi-state-like, supranational power centres that are often situated in the private sector. These entities, however, are not external to society, as they are backed and enabled by a broad, parallel alignment of users engaging in data-exposing behaviour, which the latter subjectively perceive as private, isolated and non-violent. This alignment resembles a crowd moving collectively in a certain direction, with each member believing they are acting individually, on their own account and without causing harm. In doing so, the crowd generates such momentum through its shared behaviour that it becomes unstoppable, even if individual members occasionally reflect on the potential overall consequences of their actions.

The theoretical focus on collective behaviour as ethically blameworthy, and as a basis for collective responsibility, presents a conceptual challenge: there is no inherent feature that transforms the mass of digital media users into a *collective* in the strong sense of the term. In sociology, a 'collective' typically involves a shared identity, consciousness or goal orientation. However, nothing truly collectivises us as users beyond our shared behaviour relating to our use of digital media services, coupled with the subjectivity that leads us to view these actions as individual choices. In this sense, the collective of users is not actually a *collective* at all. Instead, users are in a state of unreflective and 'serial collectivity', a term coined by Iris Marion Young, drawing on Jean-Paul Sartre, to describe gender structures (Young, 1994).[2] Serial collectivity refers to the roughly parallel alignment of users' behaviour – often unreflective – rather than a conscious phenomenon rooted in shared identities or objectives. The collective actions that need to be 'watched' and 'monitored' and the collective that must take responsibility for this 'watching' and 'monitoring' are, in fact, the same. Both have yet to be formed as a collective through the act of *taking* this responsibility by publicising and politicising the aggregate effects of a behaviour that most of us share and perceive as private in its causes and effects.

Now, the next step is to relate the state of serial collectivity of users to the informational power asymmetries that arise between industry and societies based on our shared data-exposing behaviour. As far as users are involved as enablers in AI business models, these informational power asymmetries cannot be adequately addressed as the outcome of intentional, rationally or morally explicable practices, reflections or behaviours of a group.[3] Instead, the logic through which these power asymmetries emerge from a *structural constellation* is the logic of relational, productive and strategic power operating within a sociotechnical ensemble (or *dispositif* in Foucault's sense, see Chapter 4) that loosely aligns the behaviours, perceptions and reflexivity of many individuals. In Chapter 4, I drew on Foucault's concept of subjectification (understood as the process of constituting subjectivity within a *dispositif*) to make sense of how digital media culture – embodied in artefacts, interfaces, social practices, discourses and business models – produces a parallelism between users' perceptions, practices and behaviours in such a way that subjects serve the overarching power apparatus through their individual and autonomous actions. Understanding Human–Aided AI apparatuses as cybernetic apparatuses, we can now see that the mechanisms of subjectification discussed in Chapters 3 and 4 not only make users available for exploitation within Human–Aided AI apparatuses, but they also render them unwitting *accomplices* to these apparatuses. Consequently, I argue that *this state of serial collectivity in complicity makes us collectively responsible* for the resulting effects and potential harms.

How might we describe this aspect of contemporary digital culture that aligns most users as unwitting accomplices to the power structures of prediction-based economics and governance? I believe this structural configuration is marked by a widely shared, implicit sense of entitlement to use networked services that exploit user data whenever they serve individual benefit. Not everyone perceives this behaviour as a form of entitlement and privilege – many users may instead view themselves as ignorant, having no choice or simply as individuals who 'go with the flow'. However, *structurally*, it remains an instance of entitlement, much as being racist does not require awareness. In a similar way, most people in the Global North unwittingly reproduce the 'imperial mode of living' (Brand and Wissen, 2021) with regard to resource consumption and climate change, regardless of their individual awareness and approval of, or direct contribution to, the global effects of that mode of living.

Since it is this collective manifestation of entitlement – reckless use of data-exposing Big Tech services – that a critical ethics of AI must address, *a structure-related and power-aware ethics of AI* must be formulated (see the concluding Manifesto). The aim of this ethics should be the exit of all of us, as users, from our state of serial collectivity in carelessness and our entry into a state of politicisation and responsibility-taking. As a collectivist approach, this ethics does not hold the individual consumer responsible as a solitary moral agent, but instead views *all* users and non-users as members of a responsible political community. Given the collective and structural harms caused by the use of our data, for users to exit from the state of serial collectivity calls for the activation of a sense of collectivity around responsibility. The necessary form of expression of this collective responsibility is *political action* – beginning with counter-narratives that thwart the seductive imaginaries of AI currently being foisted upon us by corporate interests and culminating in strong regulation of the data industry to address AI's power asymmetries.

Collective responsibility without responsibilisation; political action without paternalism

My critique of the sociotechnical systems of Human-Aided AI focuses on uncovering the unintentional complicity and, simultaneously, the vulnerability of each individual user within the power apparatuses of digital media. However, as I have just discussed, the transformative potential of collective responsibility lies in the mobilisation and consolidation of political – and therefore collective – counterforces. The ethical consequences of individual co-responsibility do not primarily revolve around changing one's own behaviour (such as through the conscious use or avoidance of technology), but require instead the mobilising of collectively owned regulatory programmes, designed to control and limit the manipulative and quasi-governmental power that industry players are steadily accumulating.

With my emphasis on collective responsibility, I mean to go beyond the simplistic dichotomy of consumer critique vs political regulation. Critical engagement with consumer subjectivity and habituation for the sake of political activation is a justifiable and important task, as long as this avoids the trap of fully delegating ethical responsibility to users or consumers (see my discussion of responsibilisation in Chapter 4). This pitfall is particularly evident in the political emphasis on 'behavioural change' and 'non-regulatory' government interventions based on behavioural science (for instance, nudging and other behavioural policy strategies), which are central to many conservative and liberal policy responses to, for example, the looming climate catastrophe. Behavioural change is a poor example of a consumer-focused strategy and should not be conflated with the idea of collective responsibility.[4] Such approaches shift responsibility onto individuals and their supposedly free (consumption) choices, which are cynically nudged using the same 'behavioural insights' (The House of Lords, UK, 2011) that underpin the industry's techniques of manipulation and persuasion. 'Non-regulatory' policy, therefore, leaves neoliberal ideas of the 'free market' (The House of Lords, UK, 2011) and deregulation unchallenged, and in a cowardly way evades the issue of how to curb the industry and genuinely address the accumulation of power in digital capitalism in those places where it truly occurs.

It is therefore clear that real change will only be possible through profound regulation of the Big Tech industry and the establishment of new legal frameworks that empower citizens and safeguard equitable treatment, antidiscrimination and fundamental rights in relation to how these industries operate. However, such systemic approaches to transformation come with their own challenges. Because the collective and systemic harms of aggregate data processing are relatively abstract and distant from most people's daily experience with digital services, one risk of these approaches is that they may overlook consumers as active agents, instead situating ethical and political agency solely at the macro or system level. The danger here is that individuals may be rendered passive ('politicians have to solve the problem') and individual behaviour may come to seem irrelevant ('my behaviour online makes no difference anyway'), thereby fostering feelings of powerlessness and resignation.[5]

Avoiding both extremes, the ethical approach of collective responsibility addresses individuals as citizens and members of the political community. While responsibility for change cannot be reduced to consumption choices, it *does* require individuals to participate as citizens in social and political movements that collectively demand, envision and implement structural and systemic change. The ethics of collective responsibility is, therefore, an attempt to transcend the superficial dichotomy between the neoliberal paradigm of responsibilisation and a systemic change that is being deliberated upon and implemented far above the heads of everyday consumers.

Ethics of AI as digital enlightenment

As we can see, collective responsibility requires a shared awareness of the societal harms generated by digital capitalism and enabled by our collective behaviour. This awareness needs to be channelled into a collective ethical awakening and, ultimately, into a broad political movement demanding strict regulation of the industry. Only regulation, as an expression of political will, can address the power imbalance between industry and society.

However, fostering this shared awareness as a catalyst for political change means that what Big Tech is doing to societies and individuals worldwide must *matter to us* ethically and appeal to our sense of *care for the polis*.[6] Taking responsibility in a collective sense is, interestingly, a process occurring somewhere in between the collective and the individual. On the one hand, the ethical problems of digitalisation (much like climate change) can, almost by definition, only be tackled through collective action by a majority of citizens (Hourdequin, 2010; on the debate regarding climate ethics, see Chi, 2013). On the other hand, such ethical-political mobilisation requires that a sufficient number of individuals develop an ethical and political sensibility towards what is at stake in relation to digitalisation and AI. This sensibility must ultimately be anchored in a direct existential perception on the part of the subjects. The structural lack, or even repression, of such a feeling or sensibility could be explored through a 'critical theory of the unfelt in society', as philosopher Jan Slaby puts it with respect to the climate crisis (Slaby, 2024). Activists are facing the highly complex challenge of transforming this 'societal unfeeling' (Slaby, 2024)[7] into a widely shared concern. Ultimately, this transition is a matter of the relational and collective cultivation of sensibilities and subjectivities that are perceptive to the existential dimension of what is at stake in digitalisation and AI. This is not a process that can be achieved quickly or through deliberation alone, but one that must involve education and the nurturing of ethical sensibilities as virtues of care.

Exiting the state of social unfeeling and moving towards our collective responsibility for the impacts of digital capitalism would therefore amount to something like an enlightenment movement. This is not merely a shift in public discourse on AI, but a profound transformation that involves a shared understanding of technology and business models, a critique of our habits of technology use and a heightened sensibility to the consequences for potentially anyone. Such a transition demands a collective 'will to know' (Foucault), an 'exit from self-incurred immaturity [*Unmündigkeit*]' (Kant), and the cultivation of care for the 'good life' (Aristotle) beyond 'cost-benefit analysis' and 'ethical "risk management"' (see also Vallor, 2024: 22). These concepts represent impulses that must be felt and enacted by individuals while also being collectively inspired. Though they drive public debate, they are deeply rooted in subjectivity and require a significant degree of critical self-reflection. Thus, collective ethical awareness and political action must be grounded in a spirit of enlightenment that engages people from the bottom up.

In invoking an ethics of AI in the spirit of enlightenment, I am not referring to that particular historical period's often naive belief in scientific advancement and technological progress as a solution to societal problems – an attitude echoed today in what is known as 'techno-solutionism' (Morozov, 2013). Instead, I use 'enlightenment' to signify a critical and reflective stance on power, knowledge and societal structures, as articulated by thinkers like Foucault. For Foucault, enlightenment is not a historical epoch, with its problematic social and political structures and norms (see also Mühlhoff, 2018a), but a collective event – a critical reckoning. This movement represents a break from habitualised 'non-mouthedness' (German *Unmündigkeit*, often translated as 'immaturity') and paralysis in the face of governmental techniques of paternalism and manipulation, which today are visibly practised by big tech companies. For Foucault, enlightenment is a public questioning of the present and our embeddedness within it. It marks the emergence of a specific 'type of philosophical interrogation – one which simultaneously problematizes man's [sic] relation to the present, man's historical mode of being, and the constitution of the self as an autonomous subject' within that historical moment (Foucault, 1997: 312).

An ethics of collective responsibility, grounded in the spirit of enlightenment, firmly rejects the notion that subjects are inevitably passive in the face of Big Tech. If users tend to exhibit defeatist, resigned or cynical attitudes, it is essential to analyse and understand this as a product of specific modes of subjectification within contemporary digital media culture. Defeatism presents an attitude of resignation when confronted with opaque and seemingly insurmountable forces. This may even be one of the default responses to contemporary digital capitalism, alongside cynicism – a 'smarter' form of defeatism that seeks to exploit these overpowering structures for personal gain. Ethics and critique in the digital realm must address this epoch-shaping affective and psychological constellation, not by nudging or judging users, but by activating us as citizens, recognising our ethical, critical and ultimately political agency.

Here, the task of ethics, both as a discipline and as a social movement, is to offer empowering conceptual tools, critical analyses and an inclusive discourse that supports the political quest for an enlightened attitude towards Big Tech, alongside the necessary political regulation to rein it in. Given the vast gap between individual actions and the socially harmful large-scale operations of Human-Aided AI systems, ethics remains ineffective when reduced to (moral) deliberation about right or wrong individual behaviour (see the Introduction and Wagner, 2018; Bietti, 2020; Munn, 2022; Phan et al, 2022). Much of ethics in the Western philosophical tradition, whether deontological or teleological in orientation, as well as the many applications of such classical paradigms in applied ethics of science and technology, proceeds from an implicit assumption: that there are singular individuals, called 'moral agents', who are equipped *and* motivated to engage in moral deliberation and who can, as a result of this, bring about change. However, in the face of the climate catastrophe and digital capitalism, the two most pressing societal challenges of our time, this individual-as-moral-agent paradigm leads to a dead end.

The same holds true for ethical approaches that seek to be prescriptive in terms of norms and values, or analytic in deducing moral duties from systems of assumed principles. Checklists, enumerations of 'ethical principles' and 'ethics guidelines' documents are toothless when the real problem confronting individuals and societies is a *sociotechnical complex of power* – a complex that reproduces and exploits users in their self-perceived 'helplessness', erodes the functioning of democratic publics and deadlocks societies in emerging authoritarianism, the extraction of value through technological means and the perpetuation of inequality and discrimination, all to the avail of industry interests. Given the profound subjectivity-moulding effects of digital technologies on users, what we urgently need is an ethics that addresses the *power structures* that shape our actions and perceptions – that is, an ethics of the collective patterns and alignments in subjectivity and behaviour that are not at the moral disposition of the individual, but that are co-produced by all of us as subjects within the power apparatuses (*dispositifs*) of digital consumer media.

Meeting this challenge, I suggest, is a matter of collective responsibility in a spirit of enlightenment. Such responsibility involves everyone, in that it strives to *activate* everyone to participate in a collective process of reckoning and politicising of what is at stake. Ethics in its classical, more Aristotelian sense plays a vital role in this endeavour by emphasising virtues and one's underlying moral character over isolated acts and their outcomes (Vallor, 2016, 2024). When combined with Foucault's critical enlightenment philosophy, virtue ethics becomes the search for a critical *attitude*, or *ethos*, understood as a character virtue that can spread among users, individuals and citizens. Based on analyses of power and subjectification, this approach enables us to engage in a shared reflection about how to responsibly 'hold' ourselves within the mouldings of the power structures of digital technology, of which we are inescapably a part. This ethics asks what stance we should take towards a networked technology that tends to make us all accomplices – a pull we need to resist through the attitudes we take towards this development. However, the answers to these questions – 'Which attitude?', 'What stance?' – are matters of practical wisdom (Aristotle, 2011) rather than elaborate theory; they certainly cannot be reduced to checklists or guidelines to be mechanically followed. In my framework of critical philosophy, these answers must, on the one hand, refer to something that is deeply rooted in character, as a character virtue, and on the other, be something that is shared, that exerts collective effects – engaging people, activating their inner resistance to the very power structures that shape their subjectivity and their daily habits, stimulating a shared critical reflection. As Foucault said: 'Enlightenment is not faithfulness to doctrinal elements, but rather the permanent reactivation of an attitude – that is, of a philosophical ethos that could be described as a permanent critique of our historical era' (Foucault, 1997: 312).

As a consequence, the *form of practice* of enlightenment based on critical philosophy is the systematic and methodological (both theoretical and experimental) critique of our attitude within an apparatus of power. This practice begins by asking: 'What is happening to us?', 'How does this immense power apparatus work?', 'What is our own role in this apparatus?' and 'Why do we accept it?'. The *goal* or *telos* of this practice is to overhaul and contain the power that governs us; 'the art of not being governed like that', as Foucault puts it (Foucault, 1996: 384). The foundation of this art lies in the development of a critical ethos as a character virtue. Structural and systemic changes can only be achieved as the collective manifestation of such an ethos. This is the point at which critical ethics becomes political, when individuals stop being mere users and begin to act as (global) citizens in relation to the power apparatuses of Big Tech. This shift from passivity to proactivity marks the transition from serial to active collectivity and the moment of *taking* responsibility – of caring collectively for the society of which we are all a part. No one can take this responsibility alone; the project is a collective one. The best possible outcome of scholarly work on the ethics of AI would be to support and stimulate a collective process of critical awakening and enlightened politicisation.

Conclusion: Manifesto for a Power-Aware Ethics of AI

In the face of digitalisation and AI technology, we need a new ethics. Most of AI is enabled by the collectively shared behaviour and habits of all of us as users of digital media services in the societies of the Global North and worldwide. Therefore, ethics needs to overcome its liberalist fetishising of the individual moral agent as the locus of ethical deliberation and agency. Ethics needs to abstain from the analytic habitus of judging decontextualised and simplistic scenarios. Ethics needs to become diverse. Ethics needs to become relevant and political. Ethics needs to get up from the armchair and out of the ivory tower. Ethics must go deeper than merely addressing questions of simple oughts, by cultivating character virtues that enable individuals to critically question power structures and assume societal responsibility.

What concept of AI we use is an ethical and political, not an objective question

AI is a hype. 'AI' is a catch-all container term now used to strategically label all kinds of projects that ten years ago were referred to in different and often more technical terms. Labelling things 'AI' today sells products and technologies, unlocks academic and investment funding and sparks attention. The term 'AI' is not a name referring to a more or less circumscribed part of objective reality. AI is a discursive constellation, a nexus of business and governmental interests, a conceptual vehicle that makes many people look differently, and perhaps through a more positive lens, at what in many cases could also be called 'automation'.

Any viable ethics of AI must acknowledge the political nature of the term 'AI'. Calling something AI is not a descriptive but a performative and invested act, as it creates a certain reality pertaining to the thing thus described. The political nature of the term cannot be countered by starting an ethical investigation with an objectivist 'definition of the term' that declares, once and for all, which kinds of algorithms and technologies fall into that category and which don't. In fact, these approaches are themselves no less political. They only hide their politics behind an insufficiently questioned 'culture of objectivity' (Traweek, Haraway).

We need to situate the notion of AI (and any other technology) in at least two ways. First, as sociotechnical systems marked by the mutual shaping of social reality and technological artefacts that counters naive technological determinism (Feenberg). Second, in its historicity, acknowledging that the idea of the mechanisation of intelligence has evolved as the product of power interests and hegemonic worldviews, including liberal ethical values and their implicit anthropology. Much in the ethics of AI depends on selecting a progressive and empowering understanding of AI. 'Progressive and empowering' presupposes that the concept must enable critical reflection about AI's entanglement with business models and capital interests, social inequalities and power imbalances, new forms of labour and human exploitation. 'Progressive and empowering' also presupposes that the concept must enable awareness of the intellectual and material history of that technology and thereby avoids buying into the contemporary hype surrounding AI. Finally, 'progressive and empowering' presupposes that the concept opens up new perspectives and imaginaries of AI, and inspires new voices to speak up, rather than representing a stance that merely operates as a gatekeeper that restricts the ability to exercise critical judgement and productive imagination to those who have been given the authority of 'really' knowing what they're talking about.

The ethics of AI needs to forge an alliance with social philosophy and critical theories

To take the ethics of AI where it matters most, we need to make use of the rich analytical concepts and resources developed in social and political philosophy, political economy, and science and technology studies, and critical theories including intersectional feminism, Black feminism, postcolonial studies, critical race theory and disability studies. This entails, among other things, incorporating an analysis of power structures, subjectivity and subjectification, modes of capital accumulation, forms of exploitation, subordination and abjection, and local as well as global inequalities that are inherently linked to AI technology and its reality as manifested in business and use cases.

The ethics of AI cannot be abstracted from the novel modes of value creation in digital capitalism. AI is at heart a global *industrial* development, bolstered by significant capital interests. The AI technologies and services that have tangible impacts on society and politics today cannot be detached from the IT industry in the Global North, its culture of disruptive and predatory innovation and financialised funding, its mainly White, male and ableist work culture, and its latent pioneer narratives and neocolonialist attitudes towards its own digital expansion. AI thus needs to be understood in its sociotechnical materiality that is a composite of data harvesting, user labour and user subjectivity, global inequalities and power hierarchies.

Deconstructing AI technology as an extension of historical colonialism and as an updated form of extractivism is an indispensable prerequisite for AI ethics to be relevant and not politically toothless. Any viable ethics of AI must confront AI's role in both perpetuating existing structures of discrimination, exploitation and violence, and creating new ones that do not always slot neatly into traditional categories of sexism, racism and classism. An intersectional perspective – understood as an analytic sensibility (Crenshaw) and not a formalised method – is an essential component of a critical *ethos* in AI and AI ethics. It is imperative to address the reality that discrimination, bias and opacity in AI primarily serve economic interests, uphold and maintain power differentials and drive competitive business practices (Noble). Condemning discriminatory decisions or predictions, uncovering biases and operationalising 'fairness' are now mainstream themes in AI ethics. However, these discourses backfire as bug-fixing services to the industry and the strengthening of hegemonic discourses around AI solutionism unless they confront the deep entanglements of automated injustice with economic profit, digital extraction and principles of accumulation.

The assemblage perspective on AI is not an alternative ontology but an ethical stance

We should not locate the 'intelligence' of AI systems in the innards of materially circumscribed objects like robots, chatbots or computers. We should not describe AI systems as autonomous agents or even potential moral agents (or patients), misleading the reader to believe that AI systems are entities that interact with humans similarly to how other humans would. AI systems have not popped up in our life-world as creatures that participate in human society. Instead, AI is a technology that structures our social relations, our access to and knowledge of the world and our reflexive relationship with ourselves.

To make this visible, AI systems need to be viewed as networked, decentralised and heterogeneous assemblages. However, the assemblage approach to AI should not be viewed as yet another ontological or objectivist definition of AI. The assemblage approach is not a transhumanist metaphysics; it opposes the idea that we are on an evolutionary journey from human to posthuman. Instead of a metaphysical claim, the assemblage perspective embodies a critical ethos and epistemological stance that facilitates a power-aware AI ethics. The assemblage perspective is not favoured because it is *truer*, but because it opens up a critical viewpoint. Accordingly, if someone believes in the autonomous agency of an AI system, there is no point in dismissing this belief as 'flawed'. Rather, the truth of the assemblage perspective lies in how much it enables us to question the genealogy of this belief, the discursive and material constellation that sparks it, and the power interests that benefit from it.

Thus, the point of a power-aware AI ethics in regard to anthropomorphising AI is not primarily that this stance is metaphysically *wrong* (which it is), but that it is *bad for society* in that it perpetuates the sociotechnical power constellation of the current AI hype. Morality or immorality, the good or the bad, lies in what AI technology does to our knowledge, consciousness, social lives, desires, politics and cultures. Morality inhabits the question of whose interests AI technology serves and whom it exploits, subjugates and outpowers. For example, an ambitious ethics of AI doesn't care about the 'value alignment problem' with respect to AI systems, but rather with respect to the (mostly corporate) actors building and deploying AI systems. Making those actors 'align' means regulating the industry, as a manifestation of a democratic political will.

Working in the ethics of AI requires cultivating an ethos of seeing power structures

A power-aware ethics of AI deals with structural constellations of agency that involve human and non-human factors, potentially including computing technology, capital interests, usage habits and new forms of digital labour. Humans, in general, play positive, productive and often pleasurable parts in these AI systems' operations, often unwittingly. For instance, as users and data producers, many people *want* to use specific AI devices and digital services because they subjectively benefit from them. At the same time, many people are adversely affected by those systems' wielding of power, often in indirect and not immediately apparent ways. These adversarial effects are often discernible only from an aggregate perspective as they represent harms to the political community or society as a whole, arising from serial collective behaviour that from an individual's perspective may be rational or even just fun.

Seeing and acknowledging the harmful effects of AI technology is therefore a matter of seeing *structures*. Emerging patterns of disparity, discrimination, exploitation, exclusion and misrepresentation are *aggregate* structural effects. In the liberal political mindset, as well as in the individualism of the modern Western tradition of ethics, these effects are easily overlooked and reduced to individual circumstances (as in, 'there is no gender pay gap, there are only women, each of whom individually are paid a competitive price for their work'). A power-aware ethics of AI must cultivate an ethos of seeing structures.

I use the term 'ethos' because whether or not structures exist is not primarily an ontological question but one of an onto-epistemic-ethical stance – a type of critical reasoning that is deeply rooted as a character virtue.

Recognising structures amounts to an ethical stance that refuses to blame the victims (by individualising the cause of inequality) or absolve the perpetrators (by viewing their individual contributions as marginal and failing to see the structural constellations from which they benefit).

In all of this, we need an up-to-date, dynamic understanding of 'structures' – one that avoids reverting to the static frameworks of structuralism. Structures are self-sustaining and dynamically stabilising patterns of differences, hierarchies and power relations that emerge when networks of individual microforces coalesce into supraindividual patterns of organisation. Hence, structures are emergent alignments wherein seemingly individual decisions, perspectives or behaviours collectively create a topology of power, exploitation and marginalisation that shapes society. Seeing structures is a matter of ethical and political openness to transcending the methodological individualism that has been so innate to Anglo-European science and humanities scholarship for at least a century. Seeing structures means acknowledging the collective, distributed and decentralised mechanisms that govern the emergence of patterns of inequality.

The *ethos* of seeing structures implies that an ethics of AI must not approach a specific piece of AI technology from a decontextualised, putatively objective angle that asks: 'Is this AI application good or bad?' or 'How should the system act to align with human values?'. Whose good or bad? Whose values? And who's asking? There is no uniform answer to such questions. A power-aware ethics of AI raises questions of distribution, such as: 'Who will benefit, and who won't?'; questions of power, like: 'Who will be gaining power and who will be exploited or oppressed by this technology?'; questions of participation, such as: 'Who is being heard, and who is silenced?'; and questions of economic interests, like: 'What are the business models driving this piece of innovation?'.

Ethics needs a two-step methodology that proceeds from critique to normativity

Of the three fundamental tonalities of philosophical discourse – ontological (describing what is the case), moral (prescribing what you ought to do) and critical (encouraging you to question what is taken as 'natural' in our times) – power-aware AI ethics must prioritise the critical mode of philosophical discourse. This kind of ethics must begin, as the first of two steps, with critique, rather than with morals or ontology. 'Critique' means philosophically and methodologically reflected *self-critique*. 'Self-critique' means recognising the genealogical contingency of one's own episteme (structures of thought). An important method of this kind of critical inquiry is genealogy. Genealogy seeks to tell the story of the genesis and becoming of the thought structures established in a particular discourse as shaped by power relations. In this history of becoming, certain interests have prevailed over others. As a result, it becomes apparent that its product – for example, the imaginaries of AI that are dominant today – is invested with power interests, emerging from a history of power struggles. The status of technology as objectively given is thus deconstructed by showing its dependency on material, social and political conditions. In all of this, genealogy finds its truth in its transformative effect *on subjects*, not in recording any allegedly objective history of technology.

While all this is an indispensable prerequisite, an ethics of AI must not stop at the deconstruction of universals and objective viewpoints and the inclusion of different voices and positionalities. The second step of the methodology is normativity – or *daring to be normative*, as it might feel to some scholars of a poststructuralist bent. After deconstructing the alleged naturalness and objectivity of our conception of technology, and after understanding the co-constitutive relationship between technology, society and subjectivity as invested with interests and power structures, this 'fluidified' mindset must start judging things and therefore add a bit of the moral tonality as an overtone to the critical one. After all, powerful things are happening in and to our world. Technology *is* being effective. In times of alt-right and neocolonialist narratives of entitlement, we cannot afford to be agnostic, vacillating and self-questioning forever, leaving judgement and hands-on political action to others. There are at least two things that need to be judged as loudly and visibly as possible after critical fluidification. First, we need to judge the various alternatives of naming and describing the same technological phenomenon, such as different conceptions of AI, and how useful these alternatives are in identifying unethical aspects of the status quo. Second, we need to judge existing and develop novel political proposals for the progressive regulation of technology, which means that we need to support the state in its mandate to control power imbalances.

Approaches in the ethics of AI that take a shortcut from naive objectivist conceptions of technology to normative judgement quickly exhaust themselves with irrelevant and decontextualised questions (as in the trolley problem) and the sophistical quibbling of 'armchair philosophy'. The normative and regulatory proposals from proponents of this new scholasticism will fall short of the possibilities of critical ethics because they do not understand the power dynamics that need to be addressed and contradicted. Ethics as a whole encompasses these two steps: critique first, normativity second. Importantly, step 1 cannot be outsourced from ethics to other philosophical disciplines such as critical theory. This is because the purpose of step 1 is not so much to produce a specific body of knowledge, published in papers and books, but rather to foster critically reflective individuals as philosophers. Critique as self-critique is the *personal precondition* for a sophisticated, politically relevant and power-aware ethics.

An ethics of AI must engage in political debate and avoid falling prey to 'ethics washing' and techno-fixes

Ethical questions of AI concern neither the moral behaviour of artefacts (because artefacts should not be hypostatised as moral agents) nor, in general, the moral wrongs done by specific individuals. Ethical questions of AI concern reflection, judgement and appropriate governance of the role of AI technology in the exercising of contemporary economic interests and the large-scale effects AI technology has on society. The most pertinent ethical decisions in relation to AI technology are therefore *political* decisions.

In terms of AI, the most relevant ethical questions should take the form: 'Do we *want* these actors to be allowed to create/design/apply this technology in a way that has this or that effect on us all?'. In the liberal framing of ethics in Western academic discourse over the past two centuries, considering a question like this as an ethical one has not been an obvious path to take. Moral behaviour has, in tandem with the rise of liberalism, tended to be construed as a *private* matter. There is a long way to go from where we stand today back to the ancient idea of ethics as *care for the polis* – that is, to ethics as a dimension of 'political science' (Aristotle) that is concerned with good coexistence in a political community. Ambitious ethics avoids privatisation and seeks politicisation. Insofar as it is the mandate of the democratic state to control power relations, ambitious ethics works towards a critical formation of a political will as the function of an open and inclusive public debate that leads to political participation and the forging of progressive political alliances. The site where ambitious ethics gets real is not in the moral behaviour of the individual agent, but in the political movements of collective responsibility-taking.

Active politicisation of ethical issues of AI is also much needed to evade the dominant ideology of 'solutionism'. Solutionism reifies AI technology as objectively given, and continues reifying the harmful effects of that technology as 'bugs', 'flaws' and 'glitches' that can be 'fixed' or 'ironed out' by technological improvement. Researchers who engage themselves in detecting bugs and biases or measuring the 'fairness' or otherwise of AI systems are not doing ethics of AI in the more ambitious sense; rather, they provide a welcome (free) service to the industry whose practitioners should actually be doing this work themselves. Academics drawing up ethical 'guidelines' and fashioning 'toolboxes', often commissioned by the industry, do, in most cases, take the shortcut around step 1 (critique, see 5), insofar as they're engaging in a division of labour that outsources the 'ethics component' from the development teams to 'philosophers' as contracted parties. This division of labour spares development teams the hassle of achieving the personal preconditions (see 5) for a real engagement in ethical questions. Producing corporate ethics guidelines and white papers might be a tempting way for often precariously tenured philosophers to make money, but it contributes to the hypocritical culture of 'ethics washing' that distracts from the fundamental questions of whether we, as a political community, should *want* this kind of technology – in the current way in which it is so entangled with economic and political interests – in the first place.

We must adopt a critical anthropocentrism in the ethics of AI

What is the status of the subject in an approach to AI ethics that starts from critique and tries to decentre the notion of the moral agent? After all, the goal of power-aware AI ethics is not to produce erudite tomes and papers for the library (or training data for the next large language model), but to reach human recipients in academic as well as public, political and activist debates. The reality of an ethics discourse is the *effect* it will have on thinking and feeling subjects, making them critically reflect, question the 'naturally' given, scrutinise power structures together with their peers, engage in political action and collectively take responsibility. It is thus necessary in ethics to assume that there *are* entities out there as the agents of this discourse. These entities might not be adequately captured under the monolithic notion of *the* subject, but they are human *subjects* in the sense that they are (diverse) instances of a relationship to themselves, to others and to the world. This is the consequence of a new version of an old antinomy that drives ethical work, which asserts that if there are no human recipients of the ethical discourse who *can* make a difference, who *can* make their diverse viewpoints heard, who *can* take responsibility and who *care*, then that discourse will be a fruitless and idle endeavour.

In thus assuming that there are socially and politically situated human subjects as recipients of our discourse, there is a vital quantum of (a pluralist and diversified) anthropocentrism at the heart of ethics and critical philosophy. Let's call this stance 'critical anthropocentrism'. Why does it deserve to be called critical? And why do I dare to call it anthropocentrism? First, it is critical because it starts from recognising and questioning how technology co-constitutes human consciousness, subjectivity, social relations and societal structures. It is thus not to be confused with an *ontological* theory that fixes the human as independent and separate from technology or from other species. Second, I call it anthropocentrism because it needs to be understood as a *form of discourse among entities that are susceptible to critical reckoning* – hence, human subjects in their plurality. As a philosophical stance, 'anthropocentrism' here refers to exactly the dimension of the efficacy [*Wirksamkeit*] of philosophical discourse that I hope and believe could in principle be shared with anybody on earth, but not, for instance, with a chatbot. This stance relates to the aspect of philosophical discourse that resonates with your existential status as a subject. Critical anthropocentrism is thus a philosophical stance that seeks to engage anybody for whom there *is something at stake.*

Responsibility is greater than instrumental control

Progressive AI ethics must cultivate a collective sense of responsibility that simultaneously addresses each individual, even in the light of limited agency. This means formulating a notion of responsibility that does not rely on a merely instrumentalist view of the relationship between humans and technology – one that mistakenly assumes that artefacts are just tools whose purposes can be defined and controlled by their owners. Responsibility extends far beyond what we can control instrumentally; we must take responsibility for contingent and unforeseen effects, for technological systems that we do not ourselves own, as well as for the impacts of widespread habits and unreflective practices. As many historical and contemporary examples of 'function creep' teach us, a technological invention is never confined to its primary purpose; it grows, changes and creates its own objectives and realities. Ontologically, this is a rejection of any naively instrumentalist theory of technology, while ethically it is a call for a sense of responsibility that surpasses what we can fully control.

We must therefore understand responsibility not in the forensic sense but in a proactive one – a responsibility to *intervene*, to get actively involved in the mutual shaping of technology and society. On this point, AI ethics must decisively confront a weakness of liberalism and its tendency to privatise ethics as separate from politics: AI ethics must assess the ethical quality of the social system and the political status quo as a whole. It must raise big, systemic questions: Is this the society we want to live in? Is this a good life? Do we want to be governed like that by technology and allow ourselves or others to be exploited for the interests of a few?

At the same time, and in addition to these 'big' questions, responsibility implies a deeply personal and fine-grained ethical task. While each individual only has limited agency in their direct and indirect interactions with AI technologies, we still *are* human subjects within the sociotechnical assemblages of AI. As such, we must actively show responsibility in our various roles and positions, relating our behaviour to these broader questions. In writing this book, I could only address a limited selection of relevant subject positions: as users of technology in more affluent societies, responsibility means becoming aware of the collective impact that our usage habits and data-exposing behaviour enable, as well as of this technology's dependence on the extraction of resources, data and labour across global power hierarchies. As voters, we need to recognise the urgency of strong political regulation to mitigate these effects without necessarily hindering the collective benefits of these technologies. As politicians and regulators, we must attain a solid understanding of AI and its business models if we are to be fully competent to protect our societies against the growing power asymmetries between the data industry and the people.

As ethicists, finally, we must engage in critique at the system level to be able to address the ethical issues posed by AI as questions of power, distributional inequality, exploitation and discrimination. Yet we must also avoid getting stuck in critique and deconstruction; we must be normative, and we must go public. Everyone needs to get involved, as we, as societies, must strive to gain more common knowledge about, control over and genuine political choices regarding the technologies that so fundamentally shape our presence.

Notes

Introduction

[1] I will occasionally make an empathic use of the discouraged pronoun 'we' in this book. In these moments, I am usually about to state an uncomfortable appeal or criticism. I assume that the main readership of this book is from the Global North, occupies a relatively privileged position within the global digital economy and consists of scholars in the critical humanities. When I use 'we', it is my wish to not exclude myself from this group or from any appeal to or criticism of my argument that may ensue. I am aware and fervently hope that individuals and scholars who do not exactly fit into the broad and overly simplifying category of 'Global North' (see Chapter 2, Note 5) might also read this book. To these audiences, some of my statements might still apply, though others might not.

[2] In March 2024, Google settled a lawsuit in France for breaching intellectual property rights, among others through the use of news media texts for the training of their Gemini AI system (Chrisafis, 2024).

[3] On the insufficiency of the GDPR, see Mühlhoff and Ruschemeier (2024a). Similar objections apply about the efficacy of the EU's recent AI Act with regard to limiting prediction power; see Mühlhoff and Ruschemeier (2024b).

[4] This ethical attention to the large-scale effects of AI is in line with a recent report by the *AI Now Institute*, which emphasises a material and broader conception of AI 'as a composite of data, algorithmic models, and large-scale computational power', revealing data minimisation not only as a path towards privacy protection, but also as a way to curb harmful AI applications by reducing the 'data advantage' of companies (Amba and Myers West, 2023: 10). As the report further notes: 'Data protection legislation has led to Italy's ban on ChatGPT, and Amsterdam's ruling against automated firing and algorithmic wages' (p 10).

[5] In using binary gender labels such as 'woman', 'man', 'female' and 'male' in this discussion of facial analytics software, I am following the terminology in the main source (Buolamwini and Gebru, 2018). Mapping gender onto such a binary conception is grossly reductionist and inadequate, as the authors themselves admit, but it is in alignment with industry practice. As a rule of thumb, translating 'woman' to 'femme-presenting person' (and similar for the other terms) could be a good way to go in this context.

[6] On ethics in relation to climate change, see Gardiner (2011); Cafaro (2001); Jamieson (1992). For discussions about the connection of AI and the climate crisis, see Schütze (2024b); Ullrich et al (2024).

[7] As potential starting points, see Foucault (1978); Foucault (1982); Foucault (1995); Butler (2002); Foucault (2003); Butler (2004).

[8] I occasionally use the problematic term 'Western' to designate the cultural and historical context of the Anglo-American sphere and continental Europe, particularly in relation to these regions' philosophical tradition. In these moments, I am usually about to state some kind of criticism from which I do not wish to exclude myself or my own philosophical thinking.

[9] See, for example, Anderson and Leigh (2010); Metz (2021).

[10] One might argue that the *Vienna Manifesto on Digital Humanism* aligns more closely with the approach of this book. It appears to be open to sociotechnical and even postmodern analyses of the 'complex' interplay between humans and technology, particularly in its recognition of the 'co-evolution of technology and humankind' and the way this process 'shifts power structures, thereby blurring the human and the machine' (Digital Humanism Initiative, 2019). However, in line with its universalist claims, it lacks a clear articulation and reflection of whom this collective 'we' represents. It thus misses the opportunity to stress the importance of diversity and equal participation in shaping technology for the benefit of all.

Chapter 1

[1] Paul Schütze, for example, analyses the futuristic framing of AI as an 'ideology of AI futurism' (Schütze, 2024a). Annemarie Witschas is 'demystifying the industry's mechanisms of future fabrication' by means of hegemonic AI narratives (Witschas, 2024).

[2] See also Anderson and Anderson (2011); Etzioni and Etzioni (2017); Matzner (2019).

[3] On eldercare, see Kachouie et al (2014); Hülsken-Giesler and Remmers (2020); Weber (2021). On psychotherapy, see Fiske et al (2019). On school tutoring, see Belpaeme et al (2018).

[4] For a detailed account, see also Copeland (2000); Oppy and Dowe (2021).

[5] See the historical images referenced under Figure S.1 in the supplementary material, available from: https://rainermuehlhoff.de/en/BUP-Ethics-of-AI-material.

[6] See the photo at https://www.britannica.com/topic/Deep-Blue#/media/1/155485/61084 [Accessed 10 October 2024].

[7] On the historical dimension, see Cave et al (2020).

[8] On discrimination in the context of search engines, see Noble (2018).

[9] On the use of voice analytics to detect psychic diseases, see Acosta and Weiner (2022); Skowron (2022).

[10] On credit scoring, see Lippert (2014); O'Neil (2016). On insurance risk scoring, see Kiviat (2019); Cevolini and Esposito (2020); McFall et al (2020).

[11] On personalised advertising, see Duhigg (2012); Gibbs (2015); MD Connect (2017); Andreou et al (2018); Semerádová and Weinlich (2019); Wagner and Eidenmuller (2019); Milmo (2021); Mühlhoff and Willem (2023).

[12] On political targeting, such as in the context of the Facebook/Cambridge Analytica scandal, see Sellers (2015); Grassegger and Krogerus (2016); Baldwin-Philippi (2017); Confessore and Hakim (2017); Ghoshal (2018); Hern (2018); Rosenberg et al (2018).

[13] On AI in hiring and human resource management, see Morain et al (2016); O'Neil (2016); Bogen and Rieke (2018); Raghavan and Barocas (2019); Bogen (2019); Hickman et al (2022).

[14] On predictive policing, see Angwin et al (2016); Ferguson (2016); O'Neil (2016); Dressel and Farid (2018); Martini et al (2020); Rudin et al (2020); Schwerzmann (2021). On AI in border regimes, see Chaar López (2024).

[15] On AI in the education sector, see Quinton (2015); O'Neil (2016).

[16] On AI in youth protection, see Keddell (2015); Eubanks (2017).

[17] On AI in the welfare services, see Eubanks (2017); Heikkilä (2022) DutchScandal; Ruschemeier (2023b).

Chapter 2

[1] The argument in this chapter is based on Mühlhoff (2019c); Mühlhoff (2020c); Mühlhoff (2023a).

[2] In technical terms, the idea behind this is that the system supposedly moves from a 'training' phase to an 'inference' phase. However, this distinction between training and inference is

often not present in real-world applications of machine learning methods. Even in the case of Google Image Search, the AI can never be 'fully trained' at any point, if only because new images are constantly being produced on the web, which again have to be labelled (or checked with regard to automatically generated labels) by means of human input.

3   See the screenshots in Meta Platforms, Inc. (2017) or Figure S.3 in the supplementary material, available from: https://rainermuehlhoff.de/en/BUP-Ethics-of-AI-material.

4   Following ongoing public criticism of privacy and security standards on the platform, Facebook (then rebranded as Meta Platforms, Inc.) announced in early November 2021 that it would discontinue the use of facial recognition technology to automatically tag users in photos (Facebook, 2021; Taylor, 2021).

5   Throughout the book, the terms 'Global North' and 'Global South' are utilised as shorthand to delineate global power differentials. However, it's imperative to acknowledge that these are constructed concepts for analytical expediency, rather than precise geographical designations. While these terms help to elucidate broad economic and political hierarchies, they risk oversimplifying the complexities and diversities within regions. Therefore, their usage here is intended to draw attention to systemic inequalities while recognising the need for nuanced analysis to comprehend the heterogeneous nature of global dynamics.

6   I am reusing some parts of Mühlhoff (2020c) in the presentation of this subsection. For more detailed discussions of click-work and the gig economy, see Chen (2014); Krause and Grassegger (2016); Gillespie (2018); Roberts (2019); Miceli et al (2020); Haidar and Keune (2021).

7   See, for example, the US Senate hearing with Mark Zuckerberg in April 2018, available from: https://en.wikisource.org/wiki/Zuckerberg_Senate_Transcript_2018 [Accessed 10 October 2024].

8   Google Scholar queries suggest that approximately 3,500 research articles containing the phrase 'mechanical turk' were published in the years 2008–2011, 30,800 in the years 2012–2017 and 33,000 in the years 2018–2022 (https://scholar.google.com [Accessed 2 March 2023]).

9   See also Miceli et al (2020), who discuss the power differentials in the context of the resulting computing power as arising from 'sense-making processes' or 'sense-making' labour through which 'annotators assign meaning to data through the use of labels'.

10   Acosta (2015), quoted after Gago and Mezzadra (2017: 576).

11   For a recent, extensive analysis of 'AI in and for Africa' that proceeds from critique to ethics, see also Brokensha et al (2023).

12   See Solon (2017); Malik (2022). For an extensive analysis and critique, see Coleman (2019: 428–31).

13   Couldry and Mejias' (2019) framing of the term 'data colonialism' has been heavily criticised for suggesting a qualitative shift that distances the digital versions of colonialism from the geographical contexts of historical colonialism (see Brevini et al, 2024). It is not my aim in this book to pass judgement on this important debate. As far as my argument in this book is concerned, on the one hand, the naturalisation of data worldwide as a resource, treated as freely exploitable, is an important argument that indeed extends a key aspect of historical colonialism beyond the geographical realm of historical colonialism. On the other hand, as I also argue in the text, there is a significant amount of exploitative data labour in the internet economy that is geographically situated in a way that can only be understood as a continuation of historical colonialism.

Chapter 3

1   On the politics of technical artefacts, see Winner (1980); Feenberg (2013). Specifically in relation to interface design, see Hadler and Irrgang (2015).

2   By 'genealogical' I am referring to genealogy as a specific form of critique in the tradition of Nietzsche and Foucault; see Foucault (1996); Foucault (1997); Foucault (1998b). For commentary on this, see Saar (2002); Saar (2008); Saar (2010); Koopman (2013).

3   See, for instance, the work by the 'Center for Humane Technology', founded by former Google employee Tristan Harris, available from: https://humanetech.com [Accessed 10 October 2024]. In 2018, that website was bluntly stating: 'Technology is hijacking our minds and society'. The services of Facebook, Twitter, Instagram and Google are 'part of a system designed to make us dependent'. 'What began as a race to monetization of our attention is now destroying the very foundation of our society: mental health, democracy, social relationships, and our children'.

4   As one of the late movers towards flat design, Apple started fading out skeuomorphic design around 2013; see Wingfield and Bilton (2012); Evans (2013).

5   'Pull to refresh' was invented by the US American software engineer Loren Brichter.

6   For an extension of the concept of sealed surfaces to large language models, see Lindemann (2024). Lindemann uses the notion of 'sealed knowledges' for the situation in which search engines with integrated language models deliver only singular answers to queries, which means sealing away the range of possible responses and potentially diverging sources from which the single answer is synthesised.

7   See Apple's leaked *Genius Training Student Workbook*, as reported in Biddle (2012).

Chapter 4

1   The 'soft' form of power, harnessing users as data producers, is evident in the business models and in the politics of design of AI and data companies worldwide. In this book, where I seek to establish an ethics of AI based on critique as structural critique and critique of the self, I focus on users in the Global North, as their perspective is most similar to my own status as a White, European scholar. The main thrust of this book is also to call for collective responsibility and political activation of those users who perhaps feel they neither 'suffer' from the effects of AI nor contribute to the structural harms resulting from AI.

2   On the concept of subjectivity according to Foucault, see Foucault (1982) and Foucault (1998a). On transferring the concept of subjectivity to the context of digital media, see also Breljak and Mühlhoff (2019).

3   The precursor of this argument is the notion of the 'reproduction of the conditions of production' by 'subjecting individuals to the ruling ideology' as a main function of 'ideological state apparatuses', in Althusser's formulation (see also Althusser, 1971). Althusser argued that, in addition to the economic structures of society (Marx's 'means of production'), there are ideological structures ('ideological state apparatuses') that work to maintain the dominant ideology and reproduce the conditions necessary for the existing mode of production to continue. In Althusser's eyes, these ideological state apparatuses include institutions like schools, media, religion and family; in the analysis of AI presented here, interface design, digital services and digital forms of social communication take on similar roles.

4   This is because subjectification – the process of producing or constituting subjectivity – is a function of power. As Foucault writes: 'This form of power applies itself to immediate everyday life which categorizes the individual, marks him [sic] by his own individuality, attaches him to his own identity, imposes a law of truth on him which he must recognize and which others have to recognize in him. It is a form of power which makes individuals subjects. There are two meanings of the word "subject": subject to someone else by control and dependence; and tied to his own identity by a conscience or self-knowledge. Both meanings suggest a form of power which subjugates and makes subject to' (Foucault, 1982: 781).

5   As Foucault says in *The Will to Knowledge*: '[P]ower must be understood in the first instance as the multiplicity of force relations immanent in the sphere in which they operate and which constitute their own organization; as the process which, through ceaseless struggles and confrontations, transforms, strengthens, or reverses them; as the support which these force relations find in one another, thus forming a chain or a system, or on the contrary, the disjunctions and contradictions which isolate them from one another; and lastly, as the strategies in which they take effect, whose general design or institutional crystallization is embodied in the state apparatus, in the formulation of the law, in the various social hegemonies' (Foucault, 1978: 92–3). On Foucault's understanding of power, see also (Foucault, 2003).

6   With the term 'immanent', reference is made to immanent thinking as a specific ontology, which can be traced back to the philosopher Spinoza (1985). At its core, immanence characterises being in an active form as being part of the whole (while that whole is 'nature', 'God' or 'substance'). Immanent thinking does not imply a substantial composition of that whole from independently constituted partial beings. Being as being part of the whole must be conceived of as being within, in the sense of participating in the whole that finds its expression as a whole. The basic principle of this way of thinking is therefore based on being as efficacy in the whole (*Wirken im Ganzen*; see also Mühlhoff (2018c)). That's why the being of the parts is not constitutive of the whole but immanent in it. Epistemologically, a transcendental standpoint from which to see and think about the whole must therefore be excluded. In relation to Foucault's dynamic conception of power and of the processes of subjectification, the idea of a one-sided and externally independent (non-relational) subjugation can be rejected from the standpoint of immanent thinking. Rather, the subjectification that constitutes subject(ivitie)s emerges from the myriad interplay of reciprocal interactions (or effects) of these active individuals. Thus, the term 'effect of power' is understood much less as rendering the individual passive but more importantly as active participation, while at the same time the structure of the whole manifests in this active participation. For a more detailed explanation, see Mühlhoff (2018c); Mühlhoff (2020a); also Deleuze (1988b); Deleuze (1990); Saar (2013).

7   Foucault refers to these arrangements or apparatuses as *dispositifs de pouvoir*, by which he means heterogeneous ensembles of elements constituting a whole – that is, a 'system of relations' of these elements. He uses this concept to analyse (historically) specific power apparatuses in which social, political and institutional practices, discourses of truth and subjective relations to self and others are instigated as part of a decentralised apparatus of power. The clearest statements on this dazzling concept can be found in the 1980 interview Foucault (1980). In the 1970s and 1980s, *dispositif* was often translated into English as 'apparatus'; see also Mühlhoff (2018c: 79).

8   This conceptualisation of 'subject' does not follow the Cartesian interpretation of a subject as a self-determined entity or unitary being of thinking, feeling and acting whose essence can be somehow accessed by means of self-consciousness and self-reflection. Instead, I follow a view of the subject and the self that emerges through and in power-laden or power-ridden interactions. As a consequence, the subject with all its capacities does not pre-exist power relations, but is continuously being formed and produced as an effect of the operations of power upon it and the structures of power within which the subject acts. Subjects can thus be considered as a *site* of power. For a broad overview, situating this approach in the legacy of Nietzsche, Marx and Freud, see Grosz (1990: 1). More specifically, I am following a Foucaultian approach here; for further readings on this, see Saar (2004); Saar (2010); Mühlhoff (2018c); Saar (2018); Mühlhoff (2019a).

9   On the role of resistance in Foucault's analytics of power, see Foucault (1978: 95–7); Foucault (1982: 780); Foucault (2003: 23–42).

[10] Foucault: 'Power is exercised only over free subjects, and only insofar as they are free. By this we mean individual or collective subjects who are faced with a field of possibilities in which several ways of behaving, several reactions and diverse comportments, may be realized. Where the determining factors saturate the whole, there is no relationship of power [...]. Consequently, there is no face-to-face confrontation of power and freedom, which are mutually exclusive (freedom disappears everywhere power is exercised), but a much more complicated interplay. In this game freedom may well appear as the condition for the exercise of power' (Foucault, 1982: 790).

[11] For Foucault's own interpretation of critique, see, for instance, Foucault (1996); Foucault (1997) or the lectures he gave at the Collège de France in his later years.

[12] Helpful contributions towards such a self-critical engagement come from post-Marxist theorists who have compared Facebook to a 'digital assembly line' (Scholz, 2013), or who characterise data-producing users' behaviour on the internet as 'free labour' (Terranova, 2004), 'digital labour', 'audience labour' or 'prosumer labour' (Fuchs, 2010; Fisher and Fuchs, 2015; Nixon, 2015).

[13] On the difference between sovereign or repressive forms of power and constitutive power, see Foucault (1978: 81–96); Foucault (2003: 43–62). For a wider context, see Lawlor and Nale (2014: especially 377–85 and 456–65).

Chapter 5

[1] The notion of 'frontier AI' as a reference to foundational models and general-purpose AI was originally pushed by the then UK Prime Minister Rishi Sunak in the context of the AI Safety Summit of November 2023 (attended by, among others, Elon Musk). See https://www.gov.uk/government/publications/ai-safety-summit-introduction/ai-safety-summit-introduction-html [Accessed 10 October 2024] and https://assets.publishing.service.gov.uk/media/653bc393d10f3500139a6ac5/future-risks-of-frontier-ai-annex-a.pdf [Accessed 10 October 2024].

[2] See Hildebrandt and Gutwirth (2008); O'Neil (2016); Zarsky (2016); Zuboff (2019). See also Chapter 1, Notes 8–17.

[3] I follow here the presentation in Mühlhoff (2021); Mühlhoff (2023b).

[4] See also Chapter 1, Notes 8–17.

[5] In critical debates, the principle of predictive targeting has often been discussed in the context of predictive policing and predictive criminal prosecution; see also Ferguson (2016). An important precursor to this debate is also the critique of 'predictive profiling' by Hildebrandt and colleagues; see also Hildebrandt and Gutwirth (2008). When I speak of targeted advertising and credit scoring as model cases of predictive analytics, I approach these in terms of revenue and economic relevance. Clearly, these two areas are backed with huge amounts of capital and have been the breeding grounds for predictive modelling for a long time.

[6] In the remainder of this section I am reusing and adapting some of the results jointly developed and originally published with Theresa Willem in Mühlhoff and Willem (2023).

[7] With the terminology of 'lookalike' targeting, we follow Facebook, which introduced so-called lookalike audiences in 2013. Similar mechanisms are available for other platforms, such as Google AdWords, which offers 'Customer Match', or the former Twitter, which automatically matched 'custom audiences' for its advertising customers according to their Twitter handles (Twitter, 2022). See Mühlhoff and Willem (2023).

[8] The following presentation is again based on the joint study with Theresa Willem, Mühlhoff and Willem (2023).

[9] For an example of an advertisement for a type 2 diabetes study placed on Facebook in 2018, please turn to the screenshot in Trialfacts (2018), or Figure S.4 referenced in the

supplementary material, available at https://rainermuehlhoff.de/en/BUP-Ethics-of-AI-material.

[10] See also Denecke et al (2015); MD Connect (2017); Guthrie et al (2019); Wisk et al (2019) and for a critical discussion Mühlhoff and Willem (2023).

[11] See the example in Figure S.4 of the supplementary material, available at https://rainermuehlhoff.de/en/BUP-Ethics-of-AImaterial.

[12] On the problem of secondary use of trained AI models, see Mühlhoff (2024); Mühlhoff and Ruschemeier (2024b).

Chapter 6

[1] The following list is based on research that was first published in Mühlhoff (2022); Mühlhoff and Ruschemeier (2022); Mühlhoff (2023b).

[2] See, for instance, the role photography played in the emergence of the 'right to privacy' in the US discourse (Warren and Brandeis, 1890).

[3] The following discussion is based on Mühlhoff (2021); Mühlhoff (2022); Mühlhoff (2023b).

[4] For a more detailed introduction of the concept of predictive privacy and its delineation to related concepts in privacy and data protection discourses such as 'group privacy' (see also Floridi, 2014; Helm, 2016; Taylor et al, 2016; Mittelstadt, 2017), 'inferential privacy' (Loi and Christen, 2020), 'categorical privacy' (Vedder, 1999) and 'right to reasonable inferences' (Wachter and Mittelstadt, 2019), see Mühlhoff (2021); Mühlhoff (2023b). See also Viljoen (2021) and Hong (2023) for related approaches to the problem.

[5] On predictive policing, see Ferguson (2012); Crawford and Schultz (2014). Regarding recidivism prediction, one of the currently most debated examples is the infamous Correctional Offender Management Profiling for Alternative Sanctions system by Northpointe, which is used in the US. This system has been found to be racist (Angwin et al, 2016); see Chapter 10 for a detailed discussion.

[6] Mireille Hildebrandt speaks of 'freezing the future'; see also Hildebrandt (2020).

[7] On the assessment in the legal context, see also Mühlhoff and Ruschemeier (2024a).

[8] On differential privacy in general, see Dwork (2006); for differential privacy in machine learning, see Abadi et al (2016); as combined with federated learning, see Kaissis et al (2020).

[9] This means we assume here that, for example, membership inference attacks (Shokri et al, 2017) and model inversion attacks (Fredrikson et al, 2015) are excluded.

[10] This situation is characterised by a counter-intuitive risk that is easily misunderstood: unlike what is usually assumed, an anonymised model creates the *greater* risk because in this case no regulation governs (or prevents) the reuse of the trained model.

[11] In terms of an ongoing research debate, see also Lewinski (2014); Blanke (2020); Skeba and Baumer (2020); Viljoen (2021); Mühlhoff and Ruschemeier (2022); Solove (2022); Hong (2023); Wachter (2022: 6); Mühlhoff (2024); Mühlhoff and Ruschemeier (2024a, 2024b, 2024c).

[12] A similar example is used in Mühlhoff (2024). While the example is hypothetical, it is still very realistic, as can be seen from these sources: Ma et al (2016); Acosta and Weiner (2022); Skowron (2022); Tian et al (2023).

[13] The following list is only a rough scheme that leaves out details such as data cleaning and standardisation, model testing and so on. Accounting for them separately is not important to my argument, which aims at locating the emergence of prediction power and the existing regulatory gap.

14 The precautionary principle was originally laid down in the Treaty on the Functioning of the European Union in relation to environmental protection (see 191[2] Treaty on the Functioning of the European Union). Over the last decades, the precautionary principle has increasingly been interpreted as a general principle of EU law and has also been applied in other areas (Girela, 2006).

15 I consider the use of 'open source' for publicly releasing trained models as an abuse of terminology, capitalising on the generally favourable association open source enjoys in the realm of software development. However, 'open-source software' specifically pertains to the release of source code, not the final 'compiled' program. By analogy, this would imply that both the training data and the model would need to be made available. Such an approach would have considerable ethical and data protection ramifications, as this subsection outlines.

Chapter 7

1 In this section I am following the presentation in Mühlhoff (2021).

2 By 'classical understanding of statistics' I am referring to the frequentist understanding of probability and inference, hypothesis testing and confidence intervals as a methodology that is widely taught in undergraduate classes of many empirical sciences. This approach constitutes a hegemony in 20th-century empirical science and harks back historically to the British eugenicist Ronald Fisher (1890–1962). See Joque (2022) and the following subsections.

3 See also Mühlhoff (2021: 676 sq.); Hacking (2016) for the full history of the 'logic of statistical inference', and Efron and Hastie (2018) for a discussion in relation to data science and machine learning.

4 See Karen Barad's extensive cultural analysis and criticism of these hidden assumptions in classical physics (Barad, 2007).

5 This quotation about predictive accuracy continues: 'For a data model, this translates as: fit the parameters in your model by using the data, then, using the model, predict these data and see how good the prediction is' (Breiman, 2001: 204).

Chapter 8

1 Wiener (1948a); Wiener (1948b); Ashby (1956); Mead (1968); Wiener (1989); for a critical perspective, see Hayles (2008).

2 See, for instance, Meta's Good Ideas Deserve to be Found ad campaign of 2022, promoting the value of 'personalised ads'. This campaign makes extensive use of the misleading rhetoric of 'personalisation', available from: https://www.youtube.com/watch?v=crFo wJzZOOk [Accessed 10 October 2024].

3 On the description of data and algorithms as 'performative', see also Matzner (2016); Matzner (2024: 123–30). Similar arguments that predictive systems *create* the reality they predict are made by Chun and Barnett (2021); Esposito (2022: Chapter 7); Hong (2022). The philosophical term 'performativity' that is used here has mainly been popularised by Judith Butler in the context of gender as performativity (see Butler, 1993). The term also has a longer history in speech act theory as the capacity of speech to act or to effect change in the world; see Mühlhoff (2018c: 319–23) for an introduction.

Chapter 9

1 This difference is computed by summing up the case-based profits V_i (which could be positive or negative), each multiplied by $(A_i - B_j)$, which in turn can take values −1, 0 or + 1. This number, $(A_i - B_j)$, reflects the comparison of the new with the baseline assessment

mechanism for the case *i*. The result is +1 if the new routine *A* admits a case that was rejected by the old routine *B*, 0 if both assessments match and −1 if the new model rejects a case that was admitted by the old.

[2] For a related critical reproach of AI ethics that calls for plain explainability or transparency, see Amoore (2020: 5–10).

Chapter 10

[1] In different ways, this point is made by Apprich et al (2018); Buolamwini and Gebru (2018); Chun (2018: 65); Mann and Matzner (2019); Wachter (2022).

[2] On the deep historical entanglements between technology-based mechanisms of surveillance and control and racism in the US, see Browne (2015).

[3] I first pointed out these challenges in Mühlhoff (2021). There, I also suggested variations of A/B testing (see Chapter 3) as a possible solution that might facilitate the more balanced production of feedback data in some cases – provided that the responsible agents actively want to take that path.

[4] See Chapter 1 Note 13, and see companies such as HireVue (https://www.hirevue. com/ [Accessed 10 October 2024]) and Talentech (https://talentech.com/ [Accessed 10 October 2024]) for concrete applications.

[5] On racism and heteronormativity in dating apps, see Niesen (2016); Narr (2021); Ma and Gajos (2022); Parry et al (2023).

Chapter 11

[1] The ethics we need, then, is fundamentally opposed to what in sociological theory has come to be known ever since Max Weber as 'methodological individualism'. Methodological individualism refers to the idea of explaining social phenomena by examining individual actions and the intentional states that motivate individual actors, because it is precisely from these that social phenomena would result. For example, he describes in an impressive passage how societies, when discussing social phenomena, speak of 'social collectives, such as states, associations, business corporations, foundations, as if they were individual persons' (Weber, 1968: 13). These collectives, Weber continues, 'must be treated as solely the resultants and modes of organization of the particular acts of individual persons, since these alone can be treated as agents in a course of subjectively understandable action' (Weber, 1968: 13). Seen in this way and in contrast to the view I put forward here, the whole (the social phenomenon) is the cumulative sum of individual actions (the parts). This underlines precisely why this approach is blind to structures as power effects in themselves, since structural phenomena would be reduced to individual action. In particular, this conception is blind to the emergent nature of novel power structures, such as those enabled by the rise in AI technology and predictive analytics.

[2] For a more recent commentary, see also Kear (2022), who speaks of 'serial crowds' in the context of algorithms pretty much in the sense of seriality that I explore in this chapter.

[3] Industrial actors, of course, play another role in this game – one that is certainly intentional and reflected upon. But as I am looking for a type of ethics and politicisation that could potentially lead to improved regulation of the industry, I am focusing on everyday users in this argument.

[4] For an exemplary introduction to the policy mindset of 'changing behaviour' see The House of Lords, UK (2011), particularly Chapter 5. For a systematic overview and critique, see Jones et al (2013).

[5] Also in more fundamental approaches such as Marxist and post-Marxist critique that call for class struggle and overhaul of (digital) capitalism, the role of the individual has often

been diminished, with greater emphasis put on the importance of economic and collective agents (such as classes, industries, collectives or labourers, the state, multitudes, political parties and so on) in explaining systemic transformations (Terranova, 2004; Gioia, 2019).

[6] I am referring to Aristotle with this phrase, who states that ethics is actually a part of 'political science' (Aristotle, 2011: 1095a). Aristotle does not draw the strong, liberal separation between private vs public behaviour, and indeed asserts that the political community exists prior to the individual citizen (Aristotle, 1984: 1253a18–29). See also Kraut (2022); Striker (2022).

[7] See also Slaby (2023) and the work on the notion of 'unfeeling' by Berlant (2015).

Bibliography

Abadi, M., Chu, A., Goodfellow, I., McMahan, H.B., Mironov, I., Talwar, K., et al (2016) 'Deep learning with differential privacy,' *Proceedings of the 2016 ACM SIGSAC Conference on Computer and Communications Security*, pp 308–18, doi: 10.1145/2976749.2978318.

Acosta, A. (2015) 'Nach den Plünderungen: Wege in den Post-Extraktivismus,' in Heinrich-Böll-Stiftung (ed) *Perspectivas Lateinamerika – Edition 1*, Berlin: Heinrich-Böll-Stiftung, pp 12–15, available from: https://www.boell.de/sites/default/files/perspectivas_lateinamerika_jenseits_des_raubbaus.pdf

Acosta, C.M. and Weiner, L. (2022) 'Artificial intelligence could soon diagnose illness based on the sound of your voice,' *NPR*, 10 October, available from: https://www.npr.org/2022/10/10/1127181418/ai-app-voice-diagnose-disease [Accessed 14 October 2023].

Althusser, L. (1971) 'Ideology and ideological state apparatuses (notes towards an investigation),' *Lenin and Philosophy and Other Essays*, London; New York: Monthly Review Press, pp 127–86.

Amba, K. and Myers West, S. (2023) 'AI Now 2023 landscape: confronting tech power, ' AI Now, available from: https://ainowinstitute.org/2023-landscape [Accessed 17 September 2024].

Ames, M. and Naaman, M. (29 April 2007) 'Why we tag: motivations for annotation in mobile and online media,' *Proceedings of the SIGCHI Conference on Human Factors in Computing Systems*, New York: ACM, pp 971–80, doi: 10.1145/1240624.1240772.

Amoore, L. (2020) *Cloud Ethics: Algorithms and the Attributes of Ourselves and Others*, Durham: Duke University Press.

Anderson, C. (2008) 'The end of theory: the data deluge makes the scientific method obsolete,' *Wired*, 23 June, available from: https://www.wired.com/2008/06/pb-theory/ [Accessed 21 November 2020].

Anderson, M. and Anderson, S.L. (eds) (2011) *Machine Ethics*, New York: Cambridge University Press.

Anderson, M. and Leigh, S. (2010) 'Robot be good: a call for ethical autonomous machines,' *Scientific American*, [online], available from: https://www.scientificamerican.com/article/robot-be-good/ [Accessed 23 September 2024].

Andreou, A., Venkatadri, G., Goga, O., Gummadi, K.P., Loiseau, P. and Mislove, A. (February 2018) 'Investigating ad transparency mechanisms in social media: a case study of Facebook's explanations,' in *NDSS 2018 – Network and Distributed System Security Symposium*, February 2018, pp 1–15, doi: 10.14722/ndss.2018.23204.

Andreou, A., Silva, M., Benevenuto, F., Goga, O., Loiseau, P. and Mislove, A. (February 2019) 'Measuring the Facebook advertising ecosystem,' *NDSS 2019 – Proceedings of the Network and Distributed System Security Symposium*, pp 1–15, doi: 10.14722/ndss.2019.23280.

Angwin, J., Larson, J., Mattu, S. and Kirchner, L. (2016) 'Machine bias,' ProPublica, [online], available from: https://www.propublica.org/article/machine-bias-risk-assessments-in-criminal-sentencing [Accessed 18 August 2020].

Apprich, C., Cramer, F., Chun, W.H.K. and Steyerl, H. (2018) *Pattern Discrimination*, Minneapolis; Lüneburg: University of Minnesota Press and meson press.

Arendt, H. (1963) *Eichmann in Jerusalem: A Report on the Banality of Evil*, New York: Viking Press.

Arendt, H. (1987) 'Collective responsibility,' in J.W. Bernauer (ed) *Amor Mundi: Explorations in the Faith and Thought of Hannah Arendt*, Boston: M. Nijhoff, pp 43–50.

Aristotle (1984) *The Politics*, trans C. Lord, Chicago: University of Chicago Press.

Aristotle (2011) *Nicomachean Ethics*, trans R.C. Bartlett and S.D. Collins, Chicago; London: University of Chicago Press.

Ashby, W.R. (1956) *An Introduction to Cybernetics*, London: Chapman & Hall.

Baldwin-Philippi, J. (2017) 'The myths of data-driven campaigning,' *Political Communication*, 34(4): 627–33, doi: 10.1080/10584609.2017.1372999.

Barad, K. (2007) *Meeting the Universe Halfway: Quantum Physics and the Entanglement of Matter and Meaning*, Durham: Duke University Press.

Barocas, S. and Selbst, A.D. (2016) 'Big data's disparate impact,' *California Law Review*, 104: 671.

Baruh, L. and Popescu, M. (2017) 'Big data analytics and the limits of privacy self-management,' *New Media & Society*, 19(4): 579–96, doi: 10.1177/1461444815614001.

Basu, R. (2019) 'What we epistemically owe to each other,' *Philosophical Studies*, 176(4): 915–31, doi: 10.1007/s11098-018-1219-z.

Bateson, G. (1972) *Steps to an Ecology of Mind*, Oxford: Chandler Publishing Company.

Beer, S. (1959) *Cybernetics and Management*, London: English Universities Press.

Belpaeme, T., Kennedy, J., Ramachandran, A., Scassellati, B. and Tanaka, F. (2018) 'Social robots for education: A review,' *Science Robotics*, 3(21): eaat5954, doi: 10.1126/scirobotics.aat5954.

Benjamin, R. (2019) *Race after Technology: Abolitionist Tools for the New Jim Code*, Cambridge, UK; Medford, MA: Polity.

Ben-Shahar, O. (2019) 'Data pollution,' *Journal of Legal Analysis*, 11: 104–59, doi: 10.1093/jla/laz005.

Berlant, L. (2015) 'Structures of unfeeling: mysterious skin,' *International Journal of Politics, Culture, and Society*, 28(3): 191–213, doi: 10.1007/s10767-014-9190-y.

Biddle, S. (2012) 'How to be a genius: this is Apple's secret employee training manual,' Gizmodo, [online], available from: https://gizmodo.com/how-to-be-a-genius-this-is-apples-secret-employee-trai-5938323 [Accessed 5 October 2024].

Bietti, E. (27 January 2020) 'From ethics washing to ethics bashing: a view on tech ethics from within moral philosophy,' *Proceedings of the 2020 Conference on Fairness, Accountability, and Transparency*, pp 210–9, doi: 10.1145/3351095.3372860.

Blanke, J.M. (2020) 'Protection for "inferences drawn": a comparison between the General Data Protection Regulation and the California Consumer Privacy Act,' *Global Privacy Law Review*, 1(2): 81–92.

Bogen, M. (2019) 'All the ways hiring algorithms can introduce bias,' Harvard Business Review, 6 May, available from: https://hbr.org/2019/05/all-the-ways-hiring-algorithms-can-introduce-bias [Accessed 4 March 2020].

Bogen, M. and Rieke, A. (2018) *Help Wanted: An Examination of Hiring Algorithms, Equity, and Bias*, Upturn, available from: https://apo.org.au/node/210071 [Accessed 13 March 2024].

Bohannon, J. (2015) 'Facebook will soon be able to ID you in any photo,' Science, 5 February, available from: https://www.science.org/content/article/facebook-will-soon-be-able-id-you-any-photo [Accessed 15 March 2023].

Borodovsky, J.T., Marsch, L.A. and Budney, A.J. (2018) 'Studying cannabis use behaviors with Facebook and web surveys: methods and insights,' *JMIR Public Health and Surveillance*, 4(2): e48, doi: 10.2196/publichealth.9408.

boyd, d. and Crawford, K. (2012) 'Critical questions for big data: provocations for a cultural, technological, and scholarly phenomenon,' *Information, Communication & Society*, 15(5): 662–79, doi: 10.1080/1369118X.2012.678878.

Brand, U. and Wissen, M. (2021) *The Imperial Mode of Living: Everyday Life and the Ecological Crisis of Capitalism*, London; New York: Verso Books.

Breiman, L. (2001) 'Statistical modeling: the two cultures (with comments and a rejoinder by the author),' *Statistical Science*, 16(3): 199–231, doi: 10.1214/ss/1009213726.

Breljak, A. and Mühlhoff, R. (2019) 'Was ist Sozialtheorie der Digitalen Gesellschaft? – Einleitung,' in R. Mühlhoff, A. Breljak and J. Slaby (eds) *Affekt Macht Netz: Auf dem Weg zu einer Sozialtheorie der digitalen Gesellschaft*, Bielefeld: transcript, pp 7–34, doi: 10.14361/9783837644395-001.

Brevini, B., Fubara-Manuel, I., Ludec, C.L., Jensen, J.L., Jimenez, A. and Bates, J. (2024) 'Critiques of data colonialism,' *Dialogues in Data Power*, Bristol University Press, pp 120–37, doi: 10.51952/9781529238327.ch006.

Brokensha, S., Kotzé, E. and Senekal, B.A. (2023) *AI in and for Africa: A Humanistic Perspective*, New York: Chapman and Hall/CRC, doi: 10.1201/9781003276135.

Browne, S. (2015) *Dark Matters: On the Surveillance of Blackness*, Durham: Duke University Press.

Buolamwini, J.A. (2017) *Gender Shades: Intersectional Phenotypic and Demographic Evaluation of Face Datasets and Gender Classifiers*, Thesis, Massachusetts Institute of Technology, available from: https://dspace.mit.edu/handle/1721.1/114068 [Accessed 19 September 2020].

Buolamwini, J. and Gebru, T. (21 January 2018) 'Gender shades: intersectional accuracy disparities in commercial gender classification,' *Conference on Fairness, Accountability and Transparency*, PMLR, pp 77–91, available from: http://proceedings.mlr.press/v81/buolamwini18a.html [Accessed 21 September 2020].

Burgess, M. (2023) 'ChatGPT has a big privacy problem,' *Wired*, 4 April, available from: https://www.wired.com/story/italy-ban-chatgpt-privacy-gdpr/ [Accessed 15 April 2023].

Burrell, J. (2016) 'How the machine "thinks": understanding opacity in machine learning algorithms,' *Big Data & Society*, 3(1), doi: 10.1177/2053951715622512.

Busker, T., Choenni, S. and Shoae Bargh, M. (26 September 2023) 'Stereotypes in ChatGPT: an empirical study,' *Proceedings of the 16th International Conference on Theory and Practice of Electronic Governance*, Belo Horizonte, Brazil: ACM, pp 24–32, doi: 10.1145/3614321.3614325.

Butler, J. (1993) *Bodies That Matter: On the Discursive Limits of 'Sex'*, New York: Routledge.

Butler, J. (2002) 'Bodies and power, revisited,' *Radical Philosophy*, 114: 13–19, available from: https://www.radicalphilosophy.com/article/bodies-and-power-revisited [Accessed 30 November 2021].

Butler, J. (2004) *Undoing Gender*, London; New York: Routledge.

Cafaro, P. (2001) 'Economic consumption, pleasure, and the good life,' *Journal of Social Philosophy*, 32(4): 471–86, doi: 10.1111/0047-2786.00108.

Calude, C.S. and Longo, G. (2017) 'The deluge of spurious correlations in big data,' *Foundations of Science*, 22(3): 595–612, doi: 10.1007/s10699-016-9489-4.

Cave, S., Dihal, K.S.M. and Dillon, S. (eds) (2020) *AI Narratives: A History of Imaginative Thinking about Intelligent Machines*, Oxford: Oxford University Press.

Cevolini, A. and Esposito, E. (2020) 'From pool to profile: social consequences of algorithmic prediction in insurance,' *Big Data & Society*, 7(2), doi: 10.1177/2053951720939228.

Chaar López, I. (2024) *The Cybernetic Border: Drones, Technology, and Intrusion*, Durham, London: Duke University Press.

Chen, A. (2014) 'The laborers who keep dick pics and beheadings out of your Facebook feed,' *Wired*, 23 October, available from: https://www.wired.com/2014/10/content-moderation/ [Accessed 30 January 2023].

Chi, V. (2013) 'Climate ethics: individual vs. collective responsibility and the problem of corruption,' *Stance: An International Undergraduate Philosophy Journal*, 6: 63–9, doi: 10.5840/stance201368.

Chowdhry, A. (2014) 'Facebook's DeepFace software can match faces with 97.25% accuracy,' *Forbes.com*, [online], available from: https://www.forbes.com/sites/amitchowdhry/2014/03/18/facebooks-deepface-software-can-match-faces-with-97-25-accuracy/ [Accessed 15 March 2023].

Chrisafis, A. (2024) 'Google fined €250m in France for breaching intellectual property deal,' *The Guardian*, 20 March, available from: https://www.theguardian.com/technology/2024/mar/20/google-fined-250m-euros-in-france-for-breaching-intellectual-property-rules [Accessed 30 September 2024].

Chun, W.H.K. (2017) *Updating to Remain the Same: Habitual New Media*, Cambridge, MA; London: The MIT Press.

Chun, W.H.K. (2018) 'Queering homophily,' in C. Apprich, F. Cramer, W.H.K. Chun, and H. Steyerl (eds) *Pattern Discrimination*, Minneapolis; Lüneburg: University of Minnesota Press and meson press, pp 59–97.

Chun, W.H.K. and Barnett, A.H. (2021) *Discriminating Data: Correlation, Neighborhoods, and the New Politics of Recognition*, Cambridge, MA; London: The MIT Press.

Coeckelbergh, M. (2020a) *AI Ethics*, Cambridge, MA: The MIT Press.

Coeckelbergh, M. (2020b) 'Artificial intelligence, responsibility attribution, and a relational justification of explainability,' *Science and Engineering Ethics*, 26: 2051–68, doi: 10.1007/s11948-019-00146-8.

Coleman, D. (2019) 'Digital colonialism: the 21st century scramble for Africa through the extraction and control of user data and the limitations of data protection laws,' *Michigan Journal of Race & Law*, 24: 417, doi: 10.36643/mjrl.24.2.digital.

Collins, P.H. (2000) 'Gender, Black feminism, and Black political economy,' *The ANNALS of the American Academy of Political and Social Science*, 568(1): 41–53, doi: 10.1177/000271620056800105.

Collins, P.H. (2015) 'Intersectionality's definitional dilemmas,' *Annual Review of Sociology*, 41(1): 1–20, doi: 10.1146/annurev-soc-073014-112142.

Collins, P.H. (2019) *Intersectionality as Critical Social Theory*, Durham: Duke University Press.

Collins, P.H. and Bilge, S. (2016) *Intersectionality*, Cambridge, UK; Medford, MA: Polity Press.

Confessore, N. and Hakim, D. (2017) 'Data firm says "secret sauce" aided Trump; many scoff,' *The New York Times*, 6 March, available from: https://www.nytimes.com/2017/03/06/us/politics/cambridge-analytica.html [Accessed 11 October 2022].

Cooper, K. (2023) 'OpenAI GPT-3: everything you need to know [updated],' Springboard, [online], available from: https://www.springboard.com/blog/data-science/machine-learning-gpt-3-open-ai/ [Accessed 30 September 2024].

Copeland, J. (2000) 'The Turing test,' *Minds and Machines*, 10: 519–39.

Couldry, N. and Mejias, U.A. (2019) *The Costs of Connection: How Data Is Colonizing Human Life and Appropriating It for Capitalism*, Culture and Economic Life, Stanford, CA: Stanford University Press.

Crawford, K. (2021) *Atlas of AI: Power, Politics, and the Planetary Costs of Artificial Intelligence*, New Haven: Yale University Press.

Crawford, K. and Schultz, J. (2014) 'Big data and due process: toward a framework to redress predictive privacy harms,' *Boston College Law Review*, 55: 37.

Crenshaw, K. (1989) 'Demarginalizing the intersection of race and sex: a Black feminist critique of antidiscrimination doctrine, feminist theory and antiracist politics,' *University of Chicago Legal Forum*, 1989(1): 139–67, doi: 10.1093/oso/9780198782063.003.0016.

Crenshaw, K. (1991) 'Mapping the margins: intersectionality, identity politics, and violence against women of color,' *Stanford Law Review*, 43: 1241, doi: 10.2307/1229039.

Crenshaw, K. (2015) 'Why intersectionality can't wait,' *Washington Post*, 24 September, available from: https://www.washingtonpost.com/news/in-theory/wp/2015/09/24/why-intersectionality-cant-wait/ [Accessed 19 September 2020].

D'Ignazio, C. and Klein, L.F. (2020) *Data Feminism*, Strong Ideas, Cambridge, MA: The MIT Press.

Davies, H. (2015) 'Ted Cruz using firm that harvested data on millions of unwitting Facebook users,' *The Guardian*, 11 December, available from: https://www.theguardian.com/us-news/2015/dec/11/senator-ted-cruz-president-campaign-facebook-user-data [Accessed 11 October 2022].

de Montjoye, Y.-A., Hidalgo, C.A., Verleysen, M. and Blondel, V.D. (2013) 'Unique in the crowd: the privacy bounds of human mobility,' *Scientific Reports*, 3(1): 1376, doi: 10.1038/srep01376.

Deleuze, G. (1988a) *Foucault*, trans S. Hand, Minneapolis: University of Minnesota Press.

Deleuze, G. (1988b) *Spinoza, Practical Philosophy*, trans R. Hurley, San Francisco: City Lights Books.

Deleuze, G. (1990) *Expressionism in Philosophy: Spinoza*, trans M. Joughin, New York: Zone Books.

Denecke, K., Bamidis, P., Bond, C., Gabarron, E., Househ, M., Lau, A.Y.S., et al (2015) 'Ethical issues of social media usage in healthcare,' *Yearbook of Medical Informatics*, 10(1): 137–47, doi: 10.15265/IY-2015-001.

Digital Humanism Initiative (2019) *Vienna Manifesto on Digital Humanism*, available from: https://caiml.org/dighum/dighum-manifesto/ [Accessed 9 September 2024].

Dignum, V. (2019) *Responsible Artificial Intelligence: How to Develop and Use AI in a Responsible Way*, Artificial Intelligence: Foundations, Theory, and Algorithms, Cham: Springer International Publishing, doi: 10.1007/978-3-030-30371-6.

Dornis, T.W. and Stober, S. (2024) *Urheberrecht und Training generativer KI-Modelle: Technologische und juristische Grundlagen*, Nomos, doi: 10.5771/9783748949558.

Douglas, M. (2023) 'Responsibility of OpenAI for defamation and serious invasions of privacy by ChatGPT,' *Communications Law Bulletin*, 8: 8–12.

Dressel, J. and Farid, H. (2018) 'The accuracy, fairness, and limits of predicting recidivism,' *Science Advances*, 4(1): eaao5580, doi: 10.1126/sciadv.aao5580.

Duhigg, C. (2012) 'How companies learn your secrets,' *The New York Times*, 16 February, available from: https://www.nytimes.com/2012/02/19/magazine/shopping-habits.html [Accessed 28 February 2020].

Dwork, C. (10 July 2006) 'Differential privacy,' in M. Bugliesi, B. Preneel, V. Sassone, and I. Wegener (eds) *Automata, Languages and Programming: 33rd International Colloquium, ICALP 2006, Venice, Italy*, Berlin and Heidelberg: Springer, pp 1–12.

Efron, B. and Hastie, T.J. (2018) *Computer Age Statistical Inference: Algorithms, Evidence, and Data Science*, Cambridge, UK: Cambridge University Press, doi: 10.1017/CBO9781316576533.

EPIC (2011) 'Complaint before the Federal Trade Commission: in re Facebook and the facial identification of users,' *Electronic Privacy Information Center (EPIC)*, available from: https://epic.org/documents/in-re-facebook-and-the-facial-identification-of-users/ [Accessed 15 March 2023].

Esposito, E. (2022) *Artificial Communication: How Algorithms Produce Social Intelligence*, The MIT Press, doi: 10.7551/mitpress/14189.001.0001.

Etzioni, A. and Etzioni, O. (2017) 'Incorporating ethics into artificial intelligence,' *The Journal of Ethics*, 21(4): 403–18, doi: 10.1007/s10892-017-9252-2.

Eubanks, V. (2017) *Automating Inequality: How High-Tech Tools Profile, Police, and Punish the Poor*, New York: St. Martin's Press.

Evans, C.L. (2013) 'A eulogy for skeuomorphism,' *Vice*, [online], available from: https://www.vice.com/en/article/nzzpyz/a-eulogy-for-skeuomorphism [Accessed 18 March 2024].

Everitt, B. and Skrondal, A. (2010) *The Cambridge Dictionary of Statistics*, 4th ed., Cambridge, UK; New York: Cambridge University Press.

Facebook (2010) 'Making photo tagging easier,' [online], available from: https://www.facebook.com/notes/10160198612616729/ [Accessed 15 March 2023].

Facebook (2021) 'An update on our use of face recognition,' *Meta Company Blog*, [online], available from: https://about.fb.com/news/2021/11/upd ate-on-use-of-face-recognition/ [Accessed 14 October 2022].

Feenberg, A. (2008) 'Critical theory of technology: an overview,' in G.J. Leckie and J.E. Buschman (eds) *Information Technology in Librarianship: New Critical Approaches*, Westport, CT: Libraries Unlimited, pp 31–46.

Feenberg, A. (2013) 'The mediation is the message: rationality and agency in the critical theory of technology,' *Techné: Research in Philosophy and Technology*, 17(1): 7–24, doi: 10.5840/techne20131712.

Ferguson, A.G. (2012) 'Predictive policing and reasonable suspicion,' *Emory LJ*, 62: 259.

Ferguson, A.G. (2016) 'Predictive prosecution,' *Wake Forest Law Review*, 51: 705.

Fisher, E. and Fuchs, C. (eds) (2015) *Reconsidering Value and Labour in the Digital Age*, Dynamics of Virtual Work, Houndmills, Basingstoke, Hampshire, UK; New York: Palgrave Macmillan.

Fiske, A., Henningsen, P. and Buyx, A. (2019) 'Your robot therapist will see you now: ethical implications of embodied artificial intelligence in psychiatry, psychology, and psychotherapy,' *Journal of Medical Internet Research*, 21(5): e13216, doi: 10.2196/13216.

Floridi, L. (2014) 'Open data, data protection, and group privacy,' *Philosophy & Technology*, 27(1): 1–3, doi: 10.1007/s13347-014-0157-8.

Floridi, L. (2016) 'Faultless responsibility: on the nature and allocation of moral responsibility for distributed moral actions,' *Philosophical Transactions of the Royal Society A: Mathematical, Physical and Engineering Sciences*, 374(2083): 20160112, doi: 10.1098/rsta.2016.0112.

Foucault, M. (1978) *The History of Sexuality, Volume 1: The Will to Knowledge*, trans R. Hurley, New York: Pantheon Books.

Foucault, M. (1980) 'The confession of the flesh,' *Power/Knowledge: Selected Interviews and Other Writings, 1972-1977*, New York: Pantheon Books, pp 94–228.

Foucault, M. (1982) 'The subject and power,' *Critical Inquiry*, 8(4): 777–95.

Foucault, M. (1995) *Discipline and Punish: The Birth of the Prison*, trans A. Sheridan, New York: Vintage.

Foucault, M. (1996) 'What is critique?,' in J. Schmidt (ed) *What Is Enlightenment?*, University of California Press, pp 382–98, doi: 10.1525/9780520916890-029.

Foucault, M. (1997) 'What is enlightenment?,' in P. Rabinow and J.D. Faubion (eds) *The Essential Works of Foucault, 1954-1984, Volume 1: Ethics, Subjectivity and Truth*, New York: New Press, pp 303–19.

Foucault, M. (1998a) 'Foucault (Maurice Florence),' *The Essential Works of Foucault, 1954-1984, Volume 2: Aesthetics, Method, and Epistemology*, New York: New Press, pp 459–64.

Foucault, M. (1998b) 'Nietzsche, genealogy, history,' in J.D. Faubion (ed) *Aesthetics, Method, and Epistemology: Essential Works of Foucault, 1954–1984*, New York: New Press, pp 369–91.

Foucault, M. (2003) *Society Must Be Defended: Lectures at the Collège de France, 1975-76*, New York: Picador.

Fourcade, M. and Healy, K. (2024) *The Ordinal Society*, Cambridge, MA; London: Harvard University Press.

Fredrikson, M., Jha, S. and Ristenpart, T. (12 October 2015) 'Model inversion attacks that exploit confidence information and basic countermeasures,' *Proceedings of the 22nd ACM SIGSAC Conference on Computer and Communications Security*, Denver, CO: ACM, pp 1322–33, doi: 10.1145/2810103.2813677.

Friedman, B. and Nissenbaum, H. (1996) 'Bias in computer systems,' *ACM Transactions on Information Systems (TOIS)*, 14(3): 330–47.

Fuchs, C. (2010) 'Labor in informational capitalism and on the internet,' *The Information Society*, 26(3): 179–96, doi: 10.1080/01972241003712215.

Fuchs, C. (2022) *Digital Humanism: A Philosophy for 21st Century Digital Society*, Society Now, Bingley, UK: Emerald Publishing.

Gago, V. and Mezzadra, S. (2017) 'A critique of the extractive operations of capital: toward an expanded concept of extractivism,' *Rethinking Marxism*, 29(4): 574–91, doi: 10.1080/08935696.2017.1417087.

Galloway, A.R. (2004) *Protocol: How Control Exists after Decentralization*, Leonardo, Cambridge, MA: The MIT Press.

Galloway, A.R. and Thacker, E. (2007) *The Exploit: A Theory of Networks*, Electronic Mediations 21, Minneapolis, MN: University of Minnesota Press.

Garante per la protezione dei dati personali (2023) 'Provvedimento del 30 marzo 2023 [9870832],' [online], available from: https://www.garanteprivacy.it:443/home/docweb/-/docweb-display/docweb/9870832 [Accessed 30 September 2024].

Gardiner, S.M. (2011) *A Perfect Moral Storm*, Oxford: Oxford University Press, doi: 10.1093/acprof:oso/9780195379440.001.0001.

Ghoshal, D. (2018) 'Mapped: The breathtaking global reach of Cambridge Analytica's parent company,' Quartz, [online], available from: https://qz.com/1239762/cambridge-analytica-scandal-all-the-countries-where-scl-elections-claims-to-have-worked/ [Accessed 13 February 2023].

Gibbs, S. (2015) 'Women less likely to be shown ads for high-paid jobs on Google, study shows,' *The Guardian*, 8 July, available from: https://www.theguardian.com/technology/2015/jul/08/women-less-likely-ads-high-paid-jobs-google-study [Accessed 6 February 2024].

Gillespie, T. (2018) *Custodians of the Internet: Platforms, Content Moderation, and the Hidden Decisions That Shape Social Media*, New Haven: Yale University Press.

Gioia, V. (2019) 'Individualism and social change,' *Journal of Interdisciplinary History of Ideas*, 8(16), available from: https://journals.openedition.org/jihi/286 [Accessed 5 June 2023].

Girela, M.A.R. (2006) 'Risk and reason in the European Union law,' *European Food and Feed Law Review*, 1: 270.

Gitelman, L. (ed) (2013) *'Raw Data' Is an Oxymoron*. Infrastructures, Cambridge, MA; London: The MIT Press.

Goggin, B. (2019) 'Inside Facebook's suicide algorithm: here's how the company uses artificial intelligence to predict your mental state from your posts,' *Business Insider*, 6 January, available from: https://www.businessinsider.com/facebook-is-using-ai-to-try-to-predict-if-youre-suicidal-2018-12 [Accessed 28 February 2020].

Goodfellow, I., Bengio, Y. and Courville, A. (2016) *Deep Learning*. Adaptive Computation and Machine Learning, Cambridge, MA: The MIT Press.

Grassegger, H. and Krogerus, M. (2016) 'Ich habe nur gezeigt, dass es die Bombe gibt,' *Das Magazin*, 3 December, available from: https://www.dasmagazin.ch/2016/12/03/ich-habe-nur-gezeigt-dass-es-die-bombe-gibt/ [Accessed 5 December 2016].

Gray, G.C. (2009) 'The responsibilization strategy of health and safety: neoliberalism and the reconfiguration of individual responsibility for risk,' *British Journal of Criminology*, 49(3): 326–42, doi: 10.1093/bjc/azp004.

Grosz, E.A. (1990) *Jacques Lacan: A Feminist Introduction*, London; New York: Routledge.

Gunkel, D.J. (2018) *Robot Rights*, Cambridge, MA: The MIT Press.

Gunkel, D.J. (2023) *Person, Thing, Robot: A Moral and Legal Ontology for the 21st Century and Beyond*, Cambridge, MA: The MIT Press.

Guthrie, K.A., Caan, B., Diem, S., Ensrud, K.E., Greaves, S.R., Larson, J.C., et al (2019) 'Facebook advertising for recruitment of midlife women with bothersome vaginal symptoms: a pilot study,' *Clinical Trials*, 16(5): 476–80, doi: 10.1177/1740774519846862.

Gymrek, M., McGuire, A.L., Golan, D., Halperin, E. and Erlich, Y. (2013) 'Identifying personal genomes by surname inference,' *Science*, 339(6117): 321–4, doi: 10.1126/science.1229566.

Hache, E. (2007) 'La responsabilité, une technique de gouvernementalité néolibérale?' *Raisons Politiques*, 28(4): 49, doi: 10.3917/rai.028.0049.

Hacking, I. (2016) *Logic of Statistical Inference*, London: Cambridge University Press.

Hadler, F. and Haupt, J. (2016) *Interface Critique*, Berlin: Kadmos, doi: 10.13140/RG.2.2.27453.05604.

Hadler, F. and Irrgang, D. (2015) 'Instant sensemaking, immersion and invisibility. Notes on the genealogy of interface paradigms,' *Punctum. International Journal of Semiotics*, 1(1): 7–25, doi: 10.18680/hss.2015.0002.

Haidar, J. and Keune, M. (2021) *Work and Labour Relations in Global Platform Capitalism*, Cheltenham: Edward Elgar Publishing, doi: 10.4337/9781802205138.

Halpern, O., Jagoda, P., Kirkwood, J.W. and Weatherby, L. (2022) 'Surplus data: an introduction,' *Critical Inquiry*, 48(2): 197–210, doi: 10.1086/717320.

Haraway, D. (1988) 'Situated knowledges: the science question in feminism and the privilege of partial perspective,' *Feminist Studies*, 14(3): 575–99, doi: 10.2307/3178066.

Haugeland, J. (1985) *Artificial Intelligence: The Very Idea*, Cambridge, MA: The MIT Press.

Hayles, N.K. (2005) *My Mother Was a Computer: Digital Subjects and Literary Texts*, Chicago, IL: University of Chicago Press.

Hayles, N.K. (2008) *How We Became Posthuman: Virtual Bodies in Cybernetics, Literature, and Informatics*, Chicago, IL: University of Chicago Press.

Hayles, N.K. (2016) 'Cognitive assemblages: technical agency and human interactions,' *Critical Inquiry*, 43(1): 32–55, doi: 10.1086/688293.

Heikkilä, M. (2022) 'Dutch scandal serves as a warning for Europe over risks of using algorithms,' *Politico*, [online], available from: https://www.politico.eu/article/dutch-scandal-serves-as-a-warning-for-europe-over-risks-of-using-algorithms/ [Accessed 14 January 2024].

Helm, P. (2016) 'Group privacy in times of big data. A literature review,' *Digital Culture & Society*, 2(2): 137–52, doi: 10.14361/dcs-2016-0209.

Henkel, A., Lüdtke, N., Buschmann, N. and Hochmann, L. (eds) (2018) *Reflexive Responsibilisierung: Verantwortung für nachhaltige Entwicklung*, Bielefeld, Germany: transcript, doi: 10.14361/9783839440667.

Hern, A. (2018) 'Cambridge Analytica: how did it turn clicks into votes?' *The Guardian*, 6 May, available from: https://www.theguardian.com/news/2018/may/06/cambridge-analytica-how-turn-clicks-into-votes-christopher-wylie [Accessed 11 October 2022].

Hickman, L., Bosch, N., Ng, V., Saef, R., Tay, L. and Woo, S.E. (2022) 'Automated video interview personality assessments: reliability, validity, and generalizability investigations,' *The Journal of Applied Psychology*, 107(8): 1323–51, doi: 10.1037/apl0000695.

Hildebrandt, M. (2020) 'Code-driven law: freezing the future and scaling the past,' in S. Deakin and C. Markou (eds) *Is Law Computable?: Critical Perspectives on Law and Artificial Intelligence*, Oxford: Hart Publishing, doi: 10.5040/9781509937097.

Hildebrandt, M. and Gutwirth, S. (eds) (2008) *Profiling the European Citizen: Cross-Disciplinary Perspectives*, New York: Springer.

Hofstadter, D.R. (1995) *Fluid Concepts & Creative Analogies: Computer Models of the Fundamental Mechanisms of Thought*, New York: Basic Books.

Hong, S. (2022) 'Predictions without futures,' *History and Theory*, 61(3): 371–90, doi: 10.1111/hith.12269.

Hong, S. (2023) 'Prediction as extraction of discretion,' *Big Data & Society*, 10(1), doi: 10.1177/20539517231171053.

Hourdequin, M. (2010) 'Climate, collective action and individual ethical obligations,' *Environmental Values*, 19(4): 443–64, doi: 10.3197/096327110X531552.

Hülsken-Giesler, M. and Remmers, H. (2020) *Robotische Systeme für die Pflege: Potenziale und Grenzen Autonomer Assistenzsysteme aus pflegewissenschaftlicher Sicht*, Göttingen: V&R Unipress.

Hutchinson, A. (2022) 'Meta reiterates the value of personalized ad tracking in new ad campaign,' *Social Media Today*, 5 April, available from: https://www.socialmediatoday.com/news/meta-reiterates-the-value-of-personalized-ad-tracking-in-new-ad-campaign/621642/ [Accessed 15 October 2022].

Irish, R. (2022) 'What Apple learned from skeuomorphism and why it still matters,' AppleInsider, [online], available from: https://appleinsider.com/articles/22/08/23/what-apple-learned-from-skeuomorphism-and-why-it-still-matters [Accessed 9 January 2023].

ISO (1998) 'ISO 9241-11:1998(en), Ergonomic requirements for office work with visual display terminals (VDTs) — part 11: guidance on usability,' [online], available from: https://www.iso.org/obp/ui/#iso:std:iso:9241:-11:ed-1:v1:en [Accessed 30 September 2024].

ISO (2010) 'ISO 9241-210:2010(en), Ergonomics of human-system interaction — part 210: human-centred design for interactive systems,' [online], available from: https://www.iso.org/obp/ui/#iso:std:iso:9241:-210:ed-1:v1:en [Accessed 30 September 2024].

Israel, J.I. (2001) *Radical Enlightenment: Philosophy and the Making of Modernity, 1650-1750*, Oxford; New York: Oxford University Press.

Jamieson, D. (1992) 'Ethics, public policy, and global warming,' *Science, Technology, & Human Values*, 17(2): 139–53, doi: 10.1177/016224399 201700201.

Jones, R., Pykett, J. and Whitehead, M. (2013) *Changing Behaviours: On the Rise of the Psychological State*, Cheltenham: Edward Elgar Publishing.

Joque, J. (2022) *Revolutionary Mathematics: Artificial Intelligence, Statistics and the Logic of Capitalism*, London; New York: Verso.

Kachouie, R., Sedighadeli, S., Khosla, R. and Chu, M.-T. (2014) 'Socially assistive robots in elderly care: a mixed-method systematic literature review,' *International Journal of Human-Computer Interaction*, 30(5): 369–93, doi: 10.1080/10447318.2013.873278.

Kaerlein, T. (2018) *Smartphones als digitale Nahkörpertechnologien: Zur Kybernetisierung des Alltags*, Bielefeld: Transcipt Verlag.

Kaissis, G.A., Makowski, M.R., Rückert, D. and Braren, R.F. (2020) 'Secure, privacy-preserving and federated machine learning in medical imaging,' *Nature Machine Intelligence*, 2(6): 305–11, doi: 10.1038/s42256-020-0186-1.

Kautz, H. (2022) 'The third AI summer: AAAI Robert S. Engelmore Memorial Lecture,' *AI Magazine*, 43(1): 105–25, doi: 10.1002/aaai.12036.

Kear, M. (2022) 'The moral economy of the algorithmic crowd: possessive collectivism and techno-economic rentiership,' *Competition & Change*, 26(3–4): 467–86, doi: 10.1177/1024529421990496.

Keddell, E. (2015) 'The ethics of predictive risk modelling in the Aotearoa/New Zealand child welfare context: child abuse prevention or neo-liberal tool?' *Critical Social Policy*, 35(1): 69–88, doi: 10.1177/0261018314543224.

Khan, M. and Hanna, A. (2023) 'The subjects and stages of AI dataset development: a framework for dataset accountability,' *Ohio State Technology Law Journal*, pp 171–256, doi: 10.2139/ssrn.4217148.

Kitchin, R. (2014) 'Big data, new epistemologies and paradigm shifts,' *Big Data & Society*, 1(1), doi: 10.1177/2053951714528481.

Kiviat, B. (2019) 'The moral limits of predictive practices: the case of credit-based insurance scores,' *American Sociological Review*, 84(6): 1134–58, doi: 10.1177/0003122419884917.

Koopman, C. (2013) *Genealogy as Critique: Foucault and the Problems of Modernity*, American Philosophy, Bloomington, IN: Indiana University Press.

Kosinski, M., Stillwell, D. and Graepel, T. (2013) 'Private traits and attributes are predictable from digital records of human behavior,' *Proceedings of the National Academy of Sciences*, 110(15): 5802–5, doi: 10.1073/pnas.1218772110.

Krämer, S. (2015) *Medium, Messenger, Transmission: An Approach to Media Philosophy*, Recursions: Theories of Media, Materiality, and Cultural Techniques, Amsterdam: Amsterdam University Press.

Krause, T. and Grassegger, H. (2016) 'Behind the walls of silence,' *European Press Prize*, [online], available from: https://www.europeanpressprize.com/article/stuck-web-evil-behind-walls-silence/ [Accessed 30 January 2023].

Kraut, R. (Fall 2022) 'Aristotle's ethics,' in E.N. Zalta and U. Nodelman (eds) *The Stanford Encyclopedia of Philosophy*, Metaphysics Research Lab, Stanford University.

Kröger, J.L., Lutz, O.H.-M. and Ullrich, S. (2021) 'The myth of individual control: mapping the limitations of privacy self-management,' *SSRN Electronic Journal*, doi: 10.2139/ssrn.3881776.

Kudina, O. and de Boer, B. (2024) 'Large language models, politics, and the functionalization of language,' *AI and Ethics*: pp 1–13, doi: 10.1007/s43681-024-00564-w.

Lanier, J. (2014) *Who Owns the Future?*, New York: Simon and Schuster.

Larson, J., Mattu, S., Kirchner, L. and Angwin, J. (2016) 'How we analyzed the COMPAS recidivism algorithm,' *ProPublica*, [online], available from: https://www.propublica.org/article/how-we-analyzed-the-com pas-recidivism-algorithm [Accessed 6 February 2024].

Lawlor, L. and Nale, J. (eds) (2014) *The Cambridge Foucault Lexicon*, Cambridge: Cambridge University Press, doi: 10.1017/CBO9781139022309.

Ledford, H. (2019) 'Millions of black people affected by racial bias in health-care algorithms,' *Nature*, 574(7780): 608–9, doi: 10.1038/d41586-019-03228-6.

Levitt, G.M. (2000) *The Turk, Chess Automaton*, Jefferson, NC: McFarland & Co.

Lewinski, K. von (2014) *Die Matrix des Datenschutzes Besichtigung und Ordnung eines Begriffsfeldes*, Tübingen: Mohr Siebeck.

Lin, P., Abney, K. and Jenkins, R. (eds) (2017) *Robot Ethics 2.0*, Oxford: Oxford University Press, doi: 10.1093/oso/9780190652951.001.0001.

Lindemann, N.F. (2024) 'Chatbots, search engines, and the sealing of knowledges,' *AI & Society*, doi: 10.1007/s00146-024-01944-w.

Lippert, J. (2014) 'ZestFinance issues small, high-rate loans, uses big data to weed out deadbeats,' *Washington Post*, 11 October, available from: https://www.washingtonpost.com/business/zestfinance-issues-small-high-rate-loans-uses-big-data-to-weed-out-deadbeats/2014/10/10/e34986b6-4d71-11e4-aa5e-7153e466a02d_story.html [Accessed 10 March 2020].

Lisker, M. (2023) *Von der (Un-)Möglichkeit, digital mündig zu sein: Tracking-Infrastrukturen und die Responsibilisierung des Individuums im Internet*, Thesis, Technische Universität Berlin.

Lock, S. (2022) 'What is AI chatbot phenomenon ChatGPT and could it replace humans?' *The Guardian*, 5 December, available from: https://www.theguardian.com/technology/2022/dec/05/what-is-ai-chatbot-phenome non-chatgpt-and-could-it-replace-humans [Accessed 30 September 2024].

Loi, M. and Christen, M. (2020) 'Two concepts of group privacy,' *Philosophy & Technology*, 33: 207–24, doi: 10.1007/s13347-019-00351-0.

Ma, X., Yang, H., Chen, Q., Huang, D. and Wang, Y. (16 October 2016) 'DepAudioNet: an efficient deep model for audio based depression classification,' *Proceedings of the 6th International Workshop on Audio/Visual Emotion Challenge*, New York: ACM, pp 35–42, doi: 10.1145/2988257.2988267.

Ma, Z. and Gajos, K.Z. (29 April 2022) 'Not just a preference: reducing biased decision-making on dating websites,' *Proceedings of the 2022 CHI Conference on Human Factors in Computing Systems*, New York: ACM, pp 1–14, doi: 10.1145/3491102.3517587.

Madiega, T. (2024) *Artificial Intelligence Act*. Briefing: EU Legislation in Progress PE 698.792, EPRS – European Parliamentary Research Service.

Malik, N. (2022) 'How Facebook took over the internet in Africa – and changed everything,' *The Guardian*, 20 January, available from: https://www.theguardian.com/technology/2022/jan/20/facebook-second-life-the-unstoppable-rise-of-the-tech-company-in-africa [Accessed 24 August 2023].

Mann, M. and Matzner, T. (2019) 'Challenging algorithmic profiling: the limits of data protection and anti-discrimination in responding to emergent discrimination,' *Big Data & Society*, 6(2), doi: 10.1177/2053951719895805.

Martini, M., Botta, J., Nink, D., Kolain, M. and Bertelsmann Stiftung (2020) 'Automatisch erlaubt?: Fünf Anwendungsfälle algorithmischer Systeme auf dem juristischen Prüfstand,' *Impuls Algorithmenethik*, doi: 10.11586/2019067.

Matthews, G., Deary, I.J. and Whiteman, M.C. (2003) *Personality Traits*, 2nd ed., Cambridge, UK; New York: Cambridge University Press.

Matthias, A. (2004) 'The responsibility gap: ascribing responsibility for the actions of learning automata,' *Ethics and Information Technology*, 6(3): 175–83, doi: 10.1007/s10676-004-3422-1.

Matzner, T. (2016) 'Beyond data as representation: the performativity of big data in surveillance,' *Surveillance & Society*, 14(2): 197–210, doi: 10.24908/ss.v14i2.5831.

Matzner, T. (2019) 'Autonome Trolleys und andere Probleme: Konfigurationen künstlicher Intelligenz in ethischen Debatten über selbstfahrende Kraftfahrzeuge,' *Zeitschrift für Medienwissenschaft*, 21(2): 46–55.

Matzner, T. (2024) *Algorithms: Technology, Culture, Politics*, Milton Park, Abingdon; New York: Routledge.

McFall, L., Meyers, G. and Hoyweghen, I.V. (2020) 'Editorial: the personalisation of insurance: data, behaviour and innovation,' *Big Data & Society*, 7(2), doi: 10.1177/2053951720973707.

McQuillan, D. (2022) *Resisting AI: An Anti-Fascist Approach to Artificial Intelligence*, Bristol, UK: Bristol University Press.

MD Connect (2017) *Social Media & Clinical Trial Recruitment*, [white paper], available from: https://cdn2.hubspot.net/hubfs/291282/documents/Gated_Content/White%20Paper%20-%20Clinical%20Trial%20-%20Social%20Media%20Patient%20Recruitment.pdf [Accessed 10 October 2024].

Mead, M. (1968) *Cybernetics of Cybernetics*, New York: Spartan Books.

Merchant, R.M., Asch, D.A., Crutchley, P., Ungar, L.H., Guntuku, S.C., Eichstaedt, J.C., et al (2019) 'Evaluating the predictability of medical conditions from social media posts,' *PLoS ONE*, 14(6): e0215476, doi: 10.1371/journal.pone.0215476.

Meta Platforms, Inc. (2017) 'Managing your identity on Facebook with face recognition technology,' *Meta Platforms Newsroom*, [online], available from: https://about.fb.com/news/2017/12/managing-your-identity-on-facebook-with-face-recognition-technology/ [Accessed 19 November 2024].

Metz, C. (2021) 'Can a machine learn morality?' *The New York Times*, 19 November, available from: https://www.nytimes.com/2021/11/19/technol ogy/can-a-machine-learn-morality.html [Accessed 23 September 2024].

Mezzadra, S. and Neilson, B. (2017) 'On the multiple frontiers of extraction: excavating contemporary capitalism,' *Cultural Studies*, 31(2–3): 185–204, doi: 10.1080/09502386.2017.1303425.

Miceli, M., Schuessler, M. and Yang, T. (2020) 'Between Subjectivity and Imposition: Power Dynamics in Data Annotation for Computer Vision', arXiv:2007.14886 [cs.HC] (pre-print), doi: 10.48550/arXiv.2007.14886.

Milmo, D. (2021) 'Facebook bans ads targeting race, sexual orientation and religion,' *The Guardian*, 10 November, available from: https://www.theg uardian.com/technology/2021/nov/10/facebook-bans-ads-targeting-race-sexual-orientation-and-religion [Accessed 24 March 2023].

Mittelstadt, B. (2017) 'From individual to group privacy in big data analytics,' *Philosophy & Technology*, 30(4): 475–94, doi: 10.1007/s13347-017-0253-7.

Mittelstadt, B. (2019) 'Principles alone cannot guarantee ethical AI,' *Nature Machine Intelligence*, 1(11): 501–7, doi: 10.1038/s42256-019-0114-4.

Mittelstadt, B., Allo, P., Taddeo, M., Wachter, S. and Floridi, L. (2016) 'The ethics of algorithms: mapping the debate,' *Big Data & Society*, 3(2), doi: 10.1177/2053951716679679.

Moor, J.H. (2006) 'The nature, importance, and difficulty of machine ethics,' *IEEE Intelligent Systems*, 21(4): 18–21, doi: 10.1109/MIS.2006.80.

Morain, S.R., Fowler, L.R. and Roberts, J.L. (2016) 'What to expect when [your employer suspects] you're expecting,' *JAMA Internal Medicine*, 176(11): 1597, doi: 10.1001/jamainternmed.2016.5268.

Morozov, E. (2013) *To Save Everything, Click Here: The Folly of Technological Solutionism*, New York: PublicAffairs.

Mühlhoff, R. (2018a) 'Affekte der Wahrheit. Über autoritäre Sensitivitäten von der Aufklärung bis zu 4Chan, Trump und der Alt-Right,' *Behemoth*, 11(2): 74–95, doi: 10.6094/behemoth.2018.11.2.989.

Mühlhoff, R. (2018b) 'Digitale Entmündigung und User Experience Design: Wie digitale Geräte uns nudgen, tracken und zur Unwissenheit erziehen,' *Leviathan – Journal of Social Sciences*, 46(4): 551–74, doi: 10.5771/0340-0425-2018-4-551.

Mühlhoff, R. (2018c) *Immersive Macht: Affekttheorie nach Foucault und Spinoza*, Frankfurt am Main, Germany: Campus.

Mühlhoff, R. (2019a) 'Affective disposition,' in J. Slaby and C. von Scheve (eds) *Affective Societies: Key Concepts*, New York: Routledge, pp 119–30, doi: 10.4324/9781351039260.

Mühlhoff, R. (2019b) 'Big data is watching you. Digitale Entmündigung am Beispiel von Facebook und Google,' in R. Mühlhoff, A. Breljak and J. Slaby (eds) *Affekt Macht Netz: Auf dem Weg zu einer Sozialtheorie der digitalen Gesellschaft*, Bielefeld: transcript. pp 81–107, doi: 10.14361/9783837644395-004.

Mühlhoff, R. (2019c) 'Menschengestützte Künstliche Intelligenz: Über die soziotechnischen Voraussetzungen von Deep Learning,' *Zeitschrift für Medienwissenschaft*, 21(2): 56–64, doi: 10.25969/mediarep/12633.

Mühlhoff, R. (2020a) 'Affective disposition, or: reading Spinoza with Foucault. Towards an affect theory of subjectivation and power,' *SSRN Electronic Journal*, doi: 10.2139/ssrn.3729364.

Mühlhoff, R. (2020b) 'Automatisierte Ungleichheit: Ethik der Künstlichen Intelligenz in der biopolitischen Wende des Digitalen Kapitalismus,' *Deutsche Zeitschrift für Philosophie*, 68(6): 867–90, doi: 10.1515/dzph-2020-0059.

Mühlhoff, R. (2020c) 'Human-aided artificial intelligence: or, how to run large computations in human brains? Toward a media sociology of machine learning,' *New Media & Society*, 22(10): 1868–84, doi: 10.1177/1461444819885334.

Mühlhoff, R. (2021) 'Predictive privacy: towards an applied ethics of data analytics,' *Ethics and Information Technology*, 23: 675–90, doi: 10.1007/s10676-021-09606-x.

Mühlhoff, R. (2022) 'Prädiktive Privatheit: Kollektiver Datenschutz im Kontext von Big Data und KI,' in M. Friedewald, A. Roßnagel, J. Heesen, N. Krämer and J. Lamla (eds) *Künstliche Intelligenz, Demokratie und Privatheit*, Nomos, pp 31–58, doi: 10.5771/9783748913344-31.

Mühlhoff, R. (2023a) *Die Macht der Daten: Warum künstliche Intelligenz eine Frage der Ethik ist*, Göttingen: V&R Unipress, Universitätsverlag Osnabrück, doi: 10.14220/9783737015523.

Mühlhoff, R. (2023b) 'Predictive privacy: collective data protection in times of AI and bid Data,' *Big Data & Society*, 10(1), doi: 10.1177/20539517231166886.

Mühlhoff, R. (2024) 'Das Risiko der Sekundärnutzung trainierter Modelle als zentrales Problem von Datenschutz und KI-Regulierung im Medizinbereich.' in H. Ruschemeier and B. Steinrötter (eds), *Der Einsatz von KI & Robotik in der Medizin*, Nomos, pp 27–52, doi: 10.5771/9783748939726-27.

Mühlhoff, R. and Ruschemeier, H. (2022) 'Predictive analytics und DSGVO: Ethische und rechtliche Implikationen,' in H.-C. Gräfe and Telemedicus e.V. (eds) *Telemedicus – Recht der Informationsgesellschaft, Tagungsband zur Sommerkonferenz 2022*, Frankfurt am Main, Germany: Deutscher Fachverlag, pp 38–67.

Mühlhoff, R. and Ruschemeier, H. (2024a) 'Predictive analytics and the GDPR: collective dimensions of data protection,' *Law, Innovation and Technology*, 16(1): 261–92, doi: 10.1080/17579961.2024.2313794.

Mühlhoff, R. and Ruschemeier, H. (2024b) 'Regulating AI with purpose limitation for models,' *Journal of AI Law and Regulation*, 1: 24–39, doi: 10.21552/aire/2024/1/5.

Mühlhoff, R. and Ruschemeier, H. (2024c) *Updating Purpose Limitation for AI: A Normative Approach from Law and Philosophy*, SSRN pre-print, doi: 10.2139/ssrn.4711621.

Mühlhoff, R. and Willem, T. (2023) 'Social media advertising for clinical studies: ethical and data protection implications of online targeting,' *Big Data & Society*, 10(1), doi: 10.1177/20539517231156127.

Müller, V.C. (2020) 'Ethics of artificial intelligence and robotics.' in E.N. Zalta (ed) *The Stanford Encyclopedia of Philosophy*, Stanford: Stanford University.

Munn, L. (2022) 'The uselessness of AI ethics,' *AI and Ethics*, doi: 10.1007/s43681-022-00209-w.

Narayanan, A. and Shmatikov, V. (May 2008) 'Robust de-anonymization of large sparse datasets,' *2008 IEEE Symposium on Security and Privacy (sp 2008)*, Oakland, CA; IEEE, pp 111–25, doi: 10.1109/SP.2008.33.

Narr, G. (2021) 'The uncanny swipe drive: the return of a racist mode of algorithmic thought on dating apps,' *Studies in Gender and Sexuality*, 22(3): 219–36, doi: 10.1080/15240657.2021.1961498.

Ng, A.Y. (2017) *Artificial Intelligence Is the New Electricity*, talk at Stanford Graduate School of Business, [video recording], 2 February 2017, available from: https://www.youtube.com/watch?v=21EiKfQYZXc [Accessed 14 October 2022].

Nida-Rümelin, J. and Staudacher, K. (2024) 'Philosophical foundations of digital humanism.' in H. Werthner, C. Ghezzi, J. Kramer, J. Nida-Rümelin, B. Nuseibeh, E. Prem, et al (eds) *Introduction to Digital Humanism*, Cham: Springer Nature Switzerland, pp 17–30, doi: 10.1007/978-3-031-45304-5_2.

Nida-Rümelin, J. and Weidenfeld, N. (2022) *Digital Humanism: For a Humane Transformation of Democracy, Economy and Culture in the Digital Age*, Cham: Springer International Publishing, doi: 10.1007/978-3-031-12482-2.

Niesen, M. (2016) 'Love, Inc.: toward structural intersectional analysis of online dating sites and applications,' in S.U. Noble and B.M. Tynes (eds) *The Intersectional Internet: Race, Sex, Class, and Culture Online*, New York: Peter Lang.

Nix, A. (2016) 'The power of big data and psychographics', presentation at the 2016 Concordia Annual Summit, [video recording], available from: https://www.youtube.com/watch?v=n8Dd5aVXLCc [Accessed 19 November 2024].

Nixon, B. (2015) 'The exploitation of audience labour: a missing perspective on communication and capital in the digital era,' in E. Fisher and C. Fuchs (eds) *Reconsidering Value and Labour in the Digital Age*, London: Palgrave Macmillan, pp 99–114, doi: 10.1057/9781137478573_6.

Noble, S.U. (2018) *Algorithms of Oppression: How Search Engines Reinforce Racism*, New York: University Press.

Noble, S.U. and Tynes, B.M. (eds) (2016) *The Intersectional Internet: Race, Sex, Class, and Culture Online*, New York: Peter Lang.

Norman, D.A. (1988) *The Psychology of Everyday Things*, New York: Basic Books.

Norval, A. and Prasopoulou, E. (2017) 'Public faces? A critical exploration of the diffusion of face recognition technologies in online social networks,' *New Media & Society*, 19(4): 637–54, doi: 10.1177/1461444816688896.

Nyholm, S. (2023) 'Robotic animism: the ethics of attributing minds and personality to robots with artificial intelligence,' in T. Smith (ed) *Animism and Philosophy of Religion*, Cham: Springer International Publishing, pp 313–40, doi: 10.1007/978-3-030-94170-3_13.

O'Neil, C. (2016) *Weapons of Math Destruction: How Big Data Increases Inequality and Threatens Democracy*, New York: Crown.

O'Reilly, T. (2005) 'What Is Web 2.0: Design Patterns and Business Models for the Next Generation of Software,' [online], available from: https://www.oreilly.com/lpt/a/1 [Accessed 14 March 2023].

Obermeyer, Z., Powers, B., Vogeli, C. and Mullainathan, S. (2019) 'Dissecting racial bias in an algorithm used to manage the health of populations,' *Science*, 366(6464): 447–53, doi: 10.1126/science.aax2342.

OECD.AI Policy Observatory (2019) *AI Principles Overview*, [online], available from: https://oecd.ai/en/principles [Accessed 30 September 2024].

Ohm, P. (2010) 'Broken promises of privacy: responding to the surprising failure of anonymization,' *UCLA Law Review*, 57: 1701–77.

Oksala, J. (2005) *Foucault on Freedom*, Modern European Philosophy, Cambridge, UK; New York: Cambridge University Press.

Oppy, G. and Dowe, D. (2021) 'The Turing test,' in E.N. Zalta (ed) *The Stanford Encyclopedia of Philosophy*, available from: https://plato.stanford.edu/archives/win2021/entriesuring-test/ [Accessed 20 March 2024].

Parry, D.C., Filice, E. and Johnson, C.W. (2023) 'Algorithmic heteronormativity: powers and pleasures of dating and hook-up apps,' *Sexualities*, doi: 10.1177/13634607221144626.

Pasquale, F. (2016) *The Black Box Society: The Secret Algorithms That Control Money and Information*, Cambridge, MA: Harvard University Press.

Pasquinelli, M. (2023) *The Eye of the Master: A Social History of Artificial Intelligence*, London; New York: Verso.

Paulo, N. (2023) 'The trolley problem in the ethics of autonomous vehicles,' *The Philosophical Quarterly*, 73(4): 1046–66, doi: 10.1093/pq/pqad051.

Perrigo, B. (2023) 'Exclusive: the $2 per hour workers who made ChatGPT safer,' *Time Magazine*, available from: https://time.com/6247678/openai-chatgpt-kenya-workers/ [Accessed 30 January 2023].

Peterson, M. (2019) 'The value alignment problem: a geometric approach,' *Ethics and Information Technology*, 21(1): 19–28, doi: 10.1007/s10676-018-9486-0.

Phan, T., Goldenfein, J., Mann, M. and Kuch, D. (2022) 'Economies of virtue: the circulation of "ethics" in Big Tech,' *Science as Culture*, 31(1): 121–35, doi: 10.1080/09505431.2021.1990875.

Pistilli, G. (2022) 'Debating whether AI is conscious is a distraction from real problems,' *Tech Policy Press*, [online], available from: https://techpolicy. press/debating-whether-ai-is-conscious-is-a-distraction-from-real-probl ems [Accessed 21 September 2024].

Placani, A. (2024) 'Anthropomorphism in AI: hype and fallacy,' *AI and Ethics*, 4(3): 691–8, doi: 10.1007/s43681-024-00419-4.

Proctor, D. (2018) 'Cybernetic animism: non-human personhood and the internet,' *Digital Existence*, New York: Routledge.

Pyysiäinen, J., Halpin, D. and Guilfoyle, A. (2017) 'Neoliberal governance and "responsibilization" of agents: reassessing the mechanisms of responsibility-shift in neoliberal discursive environments,' *Distinktion: Journal of Social Theory*, 18(2): 215–35, doi: 10.1080/1600910X.2017.1331858.

Quinton, S. (2015) 'Are colleges invading their students' privacy?' The Atlantic, [online], available from: https://www.theatlantic.com/educat ion/archive/2015/04/is-big-brothers-eye-on-campus/389643/ [Accessed 28 March 2020].

Raghavan, M. and Barocas, S. (2019) 'Challenges for mitigating bias in algorithmic hiring,' *Brookings*, [online], available from: https://www.brooki ngs.edu/research/challenges-for-mitigating-bias-in-algorithmic-hiring/ [Accessed 3 October 2020].

Regan, P.M. (2002) 'Privacy as a common good in the digital world,' *Information, Communication & Society*, 5(3): 382–405, doi: 10.1080/ 13691180210159328.

Rehak, R. (2021) 'The language labyrinth: constructive critique on the terminology used in the AI discourse,' in P. Verdegem (ed) *AI for Everyone? Critical Perspectives*, University of Westminster Press, pp 87–102, doi: 10.16997/book55.f.

Roberts, S. (2016a) 'Commercial content moderation: digital laborers' dirty work,' *Media Studies Publications*, available from: https://ir.lib.uwo.ca/comm pub/12 [Accessed 21 September 2024].

Roberts, S. (2016b) 'Digital refuse: Canadian garbage, commercial content moderation and the global circulation of social media's waste,' *Wi: Journal of Mobile Media*, available from: https://ir.lib.uwo.ca/commpub/14 [Accessed 21 September 2024].

Roberts, S.T. (2019) *Behind the Screen*, New Haven: Yale University Press.

Rogers, C.R. (1951) *Client Centered Therapy: Its Current Practice, Implications and Theory*, Boston: Houghton Mifflin Company.

Rosenberg, M., Confessore, N. and Cadwalladr, C. (2018) 'How Trump consultants exploited the Facebook data of millions,' *The New York Times*, 17 March, available from: https://www.nytimes.com/2018/03/17/us/politics/cambridge-analytica-trump-campaign.html [Accessed 11 October 2022].

Rouvroy, A., Athanasiadou, L. and Klumbyte, G. (2022) 'Re-imagining a "we" beyond the gathering of reductions,' *FOOTPRINT*, doi: 10.59490/FOOTPRINT.16.1.5933.

Rudin, C., Wang, C. and Coker, B. (2020) 'The age of secrecy and unfairness in recidivism prediction,' *Harvard Data Science Review*, 2(1), doi: 10.1162/99608f92.6ed64b30.

Ruschemeier, H. (2023a) 'Squaring the circle: ChatGPT and data protection,' *Verfassungsblog*, doi: 10.17176/20230407-190249-0.

Ruschemeier, H. (2023b) 'The problems of the automation bias in the public sector: a legal perspective,' in B. Berendt, M. Krzywdzinski and E. Kuznetsova (eds) *Proceedings of the Weizenbaum Conference 2023: AI, Big Data, Social Media, and People on the Move*, Weizenbaum Institute for the Networked Society, pp 59–69, doi: 10.34669/WI.CP/5.

Ruschemeier, H. (2023c) 'Data brokers and European digital legislation,' *European Data Protection Law Review*, 9(1): 27–38, doi: 10.21552/edpl/2023/1/7.

Ruschemeier, H. (2024) 'Prediction power as a challenge for the rule of law,' in O. Pollicino and P. Valcke (eds) *Oxford Handbook of Digital Constitutionalism*, Oxford, UK: Oxford University Press.

Ruschemeier, H. (2025 – forthcoming) 'Generative AI and Data Protection,' in R. Carlo, C. Poncibo, M. C., Ebers, M. and M. Zou, M. (eds), *Handbook for Generative AI and the Law*, Cambridge: Cambridge University Press.

Saar, M. (2002) 'Genealogy and subjectivity,' *European Journal of Philosophy*, 10(2): 231–45, doi: 10.1111/1468-0378.00159.

Saar, M. (2004) 'Subjekt,' in G. Göhler (ed) *Politische Theorie: 22 umkämpfte Begriffe zur Einführung*, 1st ed., UTB Politikwisssenschaft, Politische Theorie 2594, Wiesbaden, Germany: VS Verlag für Sozialwissenschaften, pp 332–49.

Saar, M. (2008) 'Understanding genealogy: history, power, and the self,' *Journal of the Philosophy of History*, 2(3): 295–314.

Saar, M. (2010) 'Power and critique,' *Journal of Power*, 3(1): 7–20.

Saar, M. (2013) *Die Immanenz der Macht: Politische Theorie nach Spinoza*, Berlin, Germany: Suhrkamp.

Saar, M. (2018) 'What is social philosophy? or: order, practice, subject,' *Proceedings of the Aristotelian Society*, 118(2): 207–23, doi: 10.1093/arisoc/aoy009.

Sag, M. (2023) 'Copyright safety for generative AI,' *Houston Law Review*, 61(2): 295–347, available from: https://houstonlawreview.org/article/92126-copyright-safety-for-generative-ai [Accessed 17 September 2024].

Samek, W., Wiegand, T. and Müller, K.-R. (2017) 'Explainable artificial intelligence: understanding, visualizing and interpreting deep learning models,' *arXiv*, doi: 10.48550/arXiv.1708.08296.

Savoldi, B., Gaido, M., Bentivogli, L., Negri, M. and Turchi, M. (2021) 'Gender bias in machine translation,' *Transactions of the Association for Computational Linguistics*, 9: 845–74, doi: 10.1162/tacl_a_00401.

Scholz, T. (ed) (2013) *Digital Labor: The Internet as Playground and Factory*, New York: Routledge.

Schütze, P. (2024a) 'The impacts of AI futurism: an unfiltered look at AI's true effects on the climate crisis,' *Ethics and Information Technology*, 26(2): 23, doi: 10.1007/s10676-024-09758-6.

Schütze, P. (2024b) 'The problem of sustainable AI: a critical assessment of an emerging phenomenon,' *Weizenbaum Journal of the Digital Society*, 4(1), doi: 10.34669/WI.WJDS/4.1.4.

Schwarcz, D. and Prince, A. (2020) 'Proxy discrimination in the age of artificial intelligence and big data,' *Iowa Law Review*, 105(3), 1257–318, available from: https://scholarship.law.umn.edu/faculty_articles/682 [Accessed 2 October 2020].

Schwerzmann, K. (2021) 'Abolish! Against the use of risk assessment algorithms at sentencing in the US criminal justice system,' *Philosophy & Technology*, 34(4): 1883–904, doi: 10.1007/s13347-021-00491-2.

Sellers, F.S. (2015) 'Cruz campaign paid $750,000 to "psychographic profiling" company,' *Washington Post*, 19 October, available from: https://www.washingtonpost.com/politics/cruz-campaign-paid-750000-to-psychographic-profiling-company/2015/10/19/6c83e508-743f-11e5-9cbb-790369643cf9_story.html [Accessed 11 October 2022].

Semerádová, T. and Weinlich, P. (2019) 'Computer estimation of customer similarity with Facebook lookalikes: advantages and disadvantages of hyper-targeting,' *IEEE Access*, 7: 153365–77, doi: 10.1109/ACCESS.2019.2948401.

Shokri, R., Stronati, M., Song, C. and Shmatikov, V. (2017) 'Membership inference attacks against machine learning models,' *arXiv*, doi: 10.48550/arXiv.1610.05820.

Skeba, P. and Baumer, E.P. (2020) 'Informational friction as a lens for studying algorithmic aspects of privacy,' *Proceedings of the ACM on Human-Computer Interaction*, 4(CSCW2): 1–22.

Skitka, L.J., Mosier, K.L. and Burdick, M. (1999) 'Does automation bias decision-making?' *International Journal of Human-Computer Studies*, 51(5): 991–1006, doi: 10.1006/ijhc.1999.0252.

Skowron, C. (2022) 'AI can use your voice to detect depression,' *Psychology Today*, [online], available from: https://www.psychologytoday.com/intl/blog/different-kind-therapy/202211/ai-can-use-your-voice-detect-depression [Accessed 10 October 2024].

Slaby, J. (2023) 'Structural apathy, affective injustice, and the ecological crisis,' *Philosophical Topics*, 51(1): 63–83.

Slaby, J. (2024) 'Habits of affluence: unfeeling, enactivism and the ecological crisis of capitalism,' *Mind & Society*, doi: 10.1007/s11299-024-00309-6.

Sloan, R.H. and Warner, R. (2014) 'Beyond notice and choice: privacy, norms, and consent,' *Journal of High Technology Law*, 14(2): 370–414, available from: https://heinonline.org/HOL/P?h=hein.journals/jhtl14&i=371 [Accessed 2 April 2022].

Smith, K.U. and Smith, M.F. (1966) *Cybernetic Principles of Learning and Educational Design*, New York: Holt, Rinehart and Winston.

Solon, O. (2017) '"It's digital colonialism": how Facebook's free internet service has failed its users,' *The Guardian*, 27 July, available from: https://www.theguardian.com/technology/2017/jul/27/facebook-free-basics-developing-markets [Accessed 18 March 2024].

Solon, O. (2018a) 'Google touts "digital wellbeing" tools to help users disengage from phones,' *The Guardian*, 8 May, available from: https://www.theguardian.com/technology/2018/may/08/google-digital-wellbeing-tools-tech-addiction [Accessed 2 October 2023].

Solon, O. (2018b) 'The rise of "pseudo-AI": how tech firms quietly use humans to do bots' work,' *The Guardian*, 6 July, available from: https://www.theguardian.com/technology/2018/jul/06/artificial-intelligence-ai-humans-bots-tech-companies [Accessed 20 July 2022].

Solove, D.J. (2022) 'The limitations of privacy rights,' *Notre Dame Law Review*, 98(3): 975–1036, doi: 10.2139/ssrn.4024790.

Spinoza, B. de (1985) 'Ethics: demonstrated in geometric order,' *The Collected Works of Spinoza, Volume I*, Annotated Edition, Princeton, NJ: Princeton University Press, pp 408–617.

Spinoza, B. de, Curley, E.M. and Spinoza, B. de (1994) *A Spinoza Reader: The Ethics and Other Works*, Princeton, NJ: Princeton University Press.

Statista (2022) 'Facebook: average revenue per user region 2021,' *Statista*, [online], available from: https://www.statista.com/statistics/251328/facebooks-average-revenue-per-user-by-region/ [Accessed 9 March 2022].

Steiger, M., Bharucha, T.J., Venkatagiri, S., Riedl, M.J. and Lease, M. (6 May 2021) 'The psychological well-being of content moderators: the emotional labor of commercial moderation and avenues for improving support,' *Proceedings of the 2021 CHI Conference on Human Factors in Computing Systems*, Yokohama, Japan: ACM, pp 1–14, doi: 10.1145/3411764.3445092.

Striker, G. (2022) 'Aristotle's ethics as political science,' in G. Striker (ed) *From Aristotle to Cicero*, Oxford: Oxford University Press, pp 128–41, doi: 10.1093/oso/9780198868385.003.0010.

Ström, T.E. (2022) 'Capital and cybernetics,' *New Left Review*, 135: 23–41.

Sweeney, L. (1997) 'Weaving technology and policy together to maintain confidentiality,' *The Journal of Law, Medicine & Ethics*, 25(2–3): 98–110, doi: 10.1111/j.1748-720X.1997.tb01885.x.

Sweeney, L. (2002) 'k-anonymity: a model for protecting privacy,' *International Journal of Uncertainty, Fuzziness and Knowledge-Based Systems*, 10(5): 557–70, doi: 10.1142/S0218488502001648.

Taddeo, M. and Floridi, L. (2018) 'How AI can be a force for good,' *Science*, 361(6404): 751–2, doi: 10.1126/science.aat5991.

Taigman, Y., Yang, M., Ranzato, M. and Wolf, L. (2014) 'DeepFace: closing the gap to human-level performance in face verification, ' *Meta Research*, available from: https://research.facebook.com/publications/deepface-clos ing-the-gap-to-human-level-performance-in-face-verification/ [Accessed 15 March 2023].

Tanwar, P. and Poply, J. (2024) *Navigating the AI IP Nexus: Legal Complexities and Forward Paths for Intellectual Property in the Age of Artificial Intelligence*, doi: 10.2139/ssrn.4804599.

Taylor, J. (2021) 'Why is Facebook shutting down its facial recognition system and deleting "faceprints"?,' *The Guardian*, 3 November, available from: https://www.theguardian.com/technology/2021/nov/03/why-is-facebook-shutting-down-its-facial-recognition-system-and-deleting-fac eprints [Accessed 14 October 2022].

Taylor, L., Floridi, L. and van der Sloot, B. (2016) *Group Privacy: New Challenges of Data Technologies*, New York: Springer.

Terranova, T. (2004) *Network Culture: Politics for the Information Age*, London; Ann Arbor, MI: Pluto Press.

Thaler, R.H. and Sunstein, C.R. (2008) *Nudge: Improving Decisions about Health, Wealth, and Happiness*, New Haven, CT: Yale University Press.

Thatcher, J., O'Sullivan, D. and Mahmoudi, D. (2016) 'Data colonialism through accumulation by dispossession: new metaphors for daily data,' *Environment and Planning D: Society and Space*, 34(6): 990–1006.

The House of Lords, UK (2011) *Science and Technology Committee – Second Report*, available from: https://publications.parliament.uk/pa/ld201012/ ldselect/ldsctech/179/17902.htm [Accessed 13 March 2024].

Tian, H., Zhu, Z. and Jing, X. (2023) 'Deep learning for depression recognition from speech,' *Mobile Networks and Applications*, doi: 10.1007/ s11036-022-02086-3.

Tiqqun (2020) *The Cybernetic Hypothesis*, trans R. Hurley, Semiotext(e).

Traweek, S. (1988) *Beamtimes and Lifetimes: The World of High Energy Physicists*, 1st ed., Cambridge, MA: Harvard University Press.

Trialfacts (2018) 'An effective clinical trial recruitment plan: narrowing the field from 500 to 24,' *Trialfacts* [online], available from: https://trialfacts. com/case-study/effective-clinical-trial-recruitment-plan-narrowing-field-from-500-to-24/ [Accessed 19 November 2024].

Tufekci, Z. (2014) 'Engineering the public: big data, surveillance and computational politics,' *First Monday*, doi: 10.5210/fm.v19i7.4901.

Turing, A. (1950) 'Computing machinery and intelligence,' *Mind*, 59(236): 433–60, doi: 10.1093/mind/LIX.236.433.

Turner, A.L. (2014) 'The history of flat design: efficiency, minimalism, trendiness,' *TNW*, [online], available from: https://thenextweb.com/news/history-flat-design-efficiency-minimalism-made-digital-world-flat [Accessed 18 March 2024].

Twitter (2022) 'Intro to custom audiences,' [online], available from: https://business.twitter.com/en/help/campaign-setup/campaign-targeting/custom-audiences.html [Accessed 18 February 2022].

Ullrich, A., Rehak, R., Hamm, A. and Mühlhoff, R. (2024) 'Editorial: sustainable artificial intelligence – critical and constructive reflections on promises and solutions, amplifications and contradictions,' *Weizenbaum Journal of the Digital Society*, 4(1), doi: 10.34669/wi.wjds/4.1.1.

Vallor, S. (2016) *Technology and the Virtues: A Philosophical Guide to a Future Worth Wanting*, New York: Oxford University Press.

Vallor, S. (2024) 'Virtues in the digital age,' in C. Véli (ed), *The Oxford Handbook of Digital Ethics*, Oxford Handbooks, Oxford; New York: Oxford University Press, pp 20–42.

van de Poel, I., Nihlén Fahlquist, J., Doorn, N., Zwart, S. and Royakkers, L. (2012) 'The problem of many hands: climate change as an example,' *Science and Engineering Ethics*, 18(1): 49–67, doi: 10.1007/s11948-011-9276-0.

van Dijck, J. (2009) 'Users like you? Theorizing agency in user-generated content,' *Media, Culture & Society*, 31(1): 41–58, doi: 10.1177/0163443708098245.

Vartan, S. (2019) 'Racial bias found in a major health care risk algorithm,' *Scientific American*, 24 October, available from: https://www.scientificamerican.com/article/racial-bias-found-in-a-major-health-care-risk-algorithm/ [Accessed 3 October 2020].

Vedder, A. (1999) 'KDD: the challenge to individualism,' *Ethics and Information Technology*, 1(4): 275–81, doi: 10.1023/A:1010016102284.

Véliz, C. (2019) 'Three things digital ethics can learn from medical ethics,' *Nature Electronics*, 2(8): 316–8, doi: 10.1038/s41928-019-0294-2.

Viljoen, S. (2021) 'A relational theory of data governance,' *Yale Law Journal*, 131(2): 573–654, available from: https://www.yalelawjournal.org/feature/a-relational-theory-of-data-governance [Accessed 3 May 2023].

von Ahn, L. (2005) *Human Computation*, Dissertation, School of Computer Science, Carnegie Mellon University.

von Ahn, L. (2006a) 'Games with a purpose,' *Computer*, 39(6): 92–4, doi: 10.1109/MC.2006.196.

von Ahn, L. (2006b) *Human Computation*, Google TechTalk, [video recording], 26 July 2006, available from: https://www.youtube.com/watch?v=tx082gDwGcM [Accessed 14 October 2022].

von Ahn, L. and Dabbish, L. (25 April 2004) 'Labeling images with a computer game,' *Proceedings of the 2004 Conference on Human Factors in Computing Systems*, Vienna, Austria: ACM Press, pp 319–26, doi: 10.1145/985692.985733.

Von Foerster, H. (1953) *Cybernetics: Circular Causal and Feedback Mechanisms in Biological and Social Systems, Transactions of the Eighth Conference, 15–16 March 1951*, New York: Josiah Macy Jr. Foundation.

W3Techs.com (2024) 'Historical trends in the usage statistics of traffic analysis tools for websites,' W3Techs, [online], available from: https://w3te chs.com/technologies/history_overview/traffic_analysis/all [Accessed 30 September 2024].

Wachter, S. (2022) 'The theory of artificial immutability: protecting algorithmic groups under anti-discrimination law,' *Tulane Law Review*, 97(2): 149–204, doi: 10.2139/ssrn.4099100.

Wachter, S. and Mittelstadt, B. (2019) 'A right to reasonable inferences: re-thinking data protection law in the age of big data and AI,' *Columbia Business Law Review*, 2019(1): 1–130, doi: 10.31228/osf.io/mu2kf.

Wagner, B. (2018) 'Ethics as an escape from regulation: from ethics-washing to ethics-shopping,' in E. Bayamlioglu, L.A.W. Janssens and M. Hildebrandt (eds) *Being Profiled: Cogitas Ergo Sum*, Amsterdam: Amsterdam University Press, pp 1–7.

Wagner, G. and Eidenmuller, H. (2019) 'Down by algorithms: siphoning rents, exploiting biases, and shaping preferences: regulating the dark side of personalized transactions,' *University of Chicago Law Review*, 86: 581.

Warren, S. and Brandeis, L. (1890) 'The right to privacy,' *Harvard Law Review*, 14(5): 193–220.

Weber, K. (2021) 'Gute Technik für ein gutes Leben?!' in D. Frommeld, U. Scorna, S. Haug and K. Weber (eds) *Gute Technik für ein gutes Leben im Alter?*, Bielefeld: transcript, pp 11–26, doi: 10.1515/9783839454695-001.

Weber, M. (1968) *Economy and Society*, G. Roth and C. Wittich (eds), Berkeley, CA: University of California Press.

Weiser, M. (1991) 'The computer for the 21st century,' *ACM SIGMOBILE*, 3: 3–11, doi: 10.1145/329124.329126.

Weizenbaum, J. (1966) 'ELIZA—a computer program for the study of natural language communication between man and machine,' *Communications of the ACM*, 9(1): 36–45, doi: 10.1145/365153.365168.

Weizenbaum, J. (1976) *Computer Power and Human Reason: From Judgment to Calculation*, San Francisco, CA: Freeman.

Whittaker, M. (2021) 'The steep cost of capture,' *Interactions*, 28(6): 50–5, doi: 10.1145/3488666.

Wiener, N. (1948a) 'Cybernetics,' *Scientific American*, 179(5): 14–19, available from: https://www.jstor.org/stable/24945913 [Accessed 8 March 2023].

Wiener, N. (1948b) *Cybernetics or Control and Communication in the Animal and the Machine.* 2nd ed., 10 print, Cambridge, MA: The MIT Press.

Wiener, N. (1989) *The Human Use of Human Beings: Cybernetics and Society,* 1954 ed., London: Free Association.

Wingfield, N. and Bilton, N. (2012) 'Apple shake-up could lead to design shift,' *The New York Times,* 1 November, available from: https://www.nytimes.com/2012/11/01/technology/apple-shake-up-could-mean-end-to-real-world-images-in-software.html [Accessed 30 September 2024].

Winner, L. (1980) 'Do artifacts have politics?' *Daedalus,* 109(1): 121–36, available from: https://www.jstor.org/stable/20024652 [Accessed 19 December 2022].

Wisk, L.E., Nelson, E.B., Magane, K.M. and Weitzman, E.R. (2019) 'Clinical trial recruitment and retention of college students with type 1 diabetes via social media: an implementation case study,' *Journal of Diabetes Science and Technology,* 13(3): 445–6, doi: 10.1177/1932296819839503.

Witschas, A. (17 June 2024) 'Prefabricated futures? AI imaginaries between elitist visions and social justice claims,' *Proceedings of the Weizenbaum Conference 2024,* Nomos.

Wong, J.C. (2021) 'Revealed: the Facebook loophole that lets world leaders deceive and harass their citizens,' *The Guardian,* 12 April, available from: https://www.theguardian.com/technology/2021/apr/12/facebook-loophole-state-backed-manipulation [Accessed 24 March 2023].

Wong, P.-H. and Simon, J. (2020) 'Thinking about "ethics" in the ethics of AI,' *Idees,* 20 February, available from: https://revistaidees.cat/en/thinking-about-ethics-in-the-ethics-of-ai/ [Accessed 4 March 2020].

Yeung, K. (2017) '"Hypernudge": big data as a mode of regulation by design,' *Information, Communication & Society,* 20(1): 118–36, doi: 10.1080/1369118X.2016.1186713.

Young, I.M. (1994) 'Gender as seriality: thinking about women as a social collective,' *Signs,* 19(3): 713–38, available from: https://www.jstor.org/stable/3174775 [Accessed 14 October 2022].

Young, I.M. (2011) *Responsibility for Justice,* New York: Oxford University Press.

Zarsky, T.Z. (2016) 'Incompatible: the GDPR in the age of big data,' *Seton Hall Law Review,* 47: 995.

Zuboff, S. (2019) *The Age of Surveillance Capitalism: The Fight for a Human Future at the New Frontier of Power,* London: Profile Books.

Index

References to figures appear in *italic* type; those in **bold** type refer to tables. References to endnotes show both the page number and the note number (187n5).